Metacognitive Regularities
and
Work Efficiency at Universities

by
Khadija Muhammad Hussain Marvi

Metacognitive Regularities
and
Work Efficiency at Universities

by
Khadija Muhammad Hussain Marvi

Academica Press
Washington

Library of Congress Cataloging-in-Publication Data

Names: Marvi, Khadija Muhammad Hussain Marvi (author)
Title: Metacognitive regularities and work efficiency at universities | Marvi, Khadija Muhammad Hussain Marvi.
Description: Washington : Academica Press, 2026. | Includes references.
Identifiers: LCCN 2026931042 | ISBN 9781680535983 (hardcover) | 9781680535990 (e-book)

Preface

This book was created to provide students with a clear and useful understanding of the core ideas of Metacognitive Regularity. I was motivated to develop this work since I observed how many learners found it difficult to make the connection between concept and Mixed Methods studies during my years of learning. The book includes key topics like work efficiency and metacognitive regularities, and each chapter is intended to deepen comprehension. The increasing need to comprehend how metacognitive regularities affect job efficiency in academic and professional contexts of faculty members at universities is the driving force behind this effort. The study aims to show how teachers can maximize their performance and productivity by looking at regular trends in planning, monitoring, and evaluation. This paper's aim is limited to finding important metacognitive regularities and examining how they relate to productivity, offering a targeted framework for enhancing productive work practices. Students and anyone else interested in learning the fundamentals of Metacognitive Regularities and Work Efficiency ought to check out this book. I hope it is a helpful study and reference tool. The Metacognitive Regularity view, which holds that people's awareness and control of cognitive processes including planning, monitoring, and evaluating follow constant regularities that affect task performance and pedagogy, serves as the foundation for this study. These Conceptual viewpoints clarify how integrated metacognitive regularities facilitate successful goal-setting, which improves control over work efficiency and increases academic productivity. I am appreciative of my teachers and responders (university teachers), who's suggestions and encouragement enabled me to complete this study.

Acknowledgement

This book is an extension of my thesis, and I am deeply indebted to those who made this journey possible. I wish to express my sincere gratitude to my advisor, Dr. Qurat ul Ain Hina, for their invaluable guidance, constructive feedback, and constant encouragement throughout my research.

I am also thankful to the faculty and staff of Educational Sciences Department/National University of Modern Languages Islamabad, whose support and resources greatly contributed to this work. Special thanks to Dr. Khushbakht Hina, Dr. Jamila Begum and Dr. Shazia Zameer for their insightful suggestions that shaped this study into its present form, And more deeply, academic teachers who completed my education Dr. Wajeeha Shahid, Dr. Marium Din, Dr. Saira Nudrat, and Dr. Farkhanda Tabassum.

Lastly, I owe heartfelt thanks to my family for their unwavering support and patience, without which this achievement would not have been possible.

Dedicated to

To my grandparents, whose wisdom have laid the foundation for all that I strive to achieve and to my beloved sons, whose presence in my life is a constant source of motivation and joy.

Table of Contents

List of Tables

List of Figures

List of Abbreviations

Abbrcviation	**Terms**
BERA	British Educational Research Association
DL	Digital
ELER	Easily learnt, easily remembered
IT	Informational Technology
MRS	Metacognitive Regulatory Scale
MRAS	Metacognitive Regularities Assessment Scale
NUML	National University of Modern Languages
RMF	Retrieval-monitoring feedback
TMR	Teachers Metacognitive Regularities
SPSS	Statistical Package for Social Sciences
WES	Work Efficiency Scale
WEAS	Work Efficiency Assessment Scale

Chapter One

Introduction

1.1. Background of the Study

Metacognitive regularity is considered a key component for the future by 2030 (Organization for Economic Co-operation & Development, 2019). Teachers can collaborate with university administrative bodies to collectively shape students' futures. Teaching professionals, such as university teachers, are best suited to utilize metacognitive regularities in their practices. Work efficiency is an essential aspect of teachers' professional lives. Teaching constitutes the academic practice of university teachers and metacognitive regularities play a key role in establishing excellence in their work. These metacognitive regulations should be implemented alongside the rules that govern teachers' professional conduct. In every era, a new generation acquires knowledge and skills through the teachings of the previous generation, and, in turn, passes on its experience to the next. According to Sergey, Arkhipov, Vanchikova, Zolotareva, Yantranov, and Budaeva, (2019) stated that in practice, classroom management, and years of teaching experience are reflected in students' development. The impact of teachers' work is clearly evident within the university system. Additionally, Siti, Sharifah, and Nik (2012) stated that modern teachers consistently focus on developing the cognitive ability to embrace and adapt to new educational concepts through their work practices. Anne, Tara, and Linda (2019) stated that the work of 21st-century teachers differs greatly from that of the past; they use a range of modern technologies to implement the latest pedagogies, making lessons engaging and effective for learners. The inspiring personalities of university teachers make this possible for students.

Fulfilling responsibilities, carefully planning, evaluating, and monitoring while staying attuned to intellectual challenges leads to work efficiency. For university teachers, responsibility in their work relies on precise monitoring, which helps ensure duties are remembered and fulfilled (Ilona & Sarka, 2014). According to personal responsibility, Jalal (2016) states that committed teachers are recognized for their efficiency in meeting their professional responsibilities. Moreover, Valerie, Robert, and Julia (1991) stated that Teachers focus on fundamental student learning and work efficacy, but a lack of regularity leads to low efficiency. However, metacognitive regularity can help establish and maintain efficiency.

Work efficiency evaluation is conducted when both information and tasks are assessed collectively. Although they differ in purpose and content, a university teacher must perform both simultaneously. The work efficiency of university teachers can be improved through the evaluation of their teaching (Gloria, 2005). Therefore, Fiona (2003) stated that a teacher can manage university work efficiently by planning tasks ahead of time. Likewise, in the classroom, teachers who plan before teaching are more focused on their students and more attentive in their pedagogical approach. Teachers' effectiveness is linked to the fact that planning takes time but ultimately enhances work efficiency.

Metacognition was traditionally considered a single cognitive component; however, in recent times, metacognitive regularities (activities that help regulate one's work) have been recognized as contributing to work efficiency. This can help teachers implement beneficial improvements in their practice. Therefore, it is stated that teachers play a vital role in accomplishing university tasks efficiently, thereby promoting student and institutional sustainability (Tahseen, Ouyang, Muhammad, Jordi, Sumaira, & Noor, 2020).

> Teachers face complex challenges that require not only subject expertise but also a high level of excellence (Ana & Mihaela, 2018). The subject of this study, metacognitive regulation, helps in organizing teachers' work to enhance efficiency. The purpose was to investigate the metacognitive regularities of university teachers, determine their work efficiency, and assess the effect of metacognitive regularities on that efficiency.

Metacognitive regularities have been widely studied in the context of student learning; however, their application in teaching practices remains largely underexplored (Julie, Amanda, & John, 2021). The concept of metacognitive regularities is relatively new for university teachers, and it is discussed further in the next chapter to support the rationale and problem statement with relevant theories and models. This discussion will elaborate on its importance, reflect on the quality of performance, and analyze related research in this field.

a. Objectives

1. To investigate the metacognitive regularities of university teachers.

2. To determine the work efficiency of university teachers.

3. To assess the effect of metacognitive regularities on work efficiency of university teachers.

3a. To assess the effect of planning on work efficiency of university teachers.

3b. To assess the effect of monitoring on work efficiency of university teachers.

3c. To assess the effect of evaluation on work efficiency of university teachers.

b. Research Questions

1. What are the metacognitive regularities of university teachers?

2. What is the level of the work efficiency of university teachers?

3. How do university teachers consider the effect of their metacognitive regularities on their work efficiency?

c. Null Hypotheses

H_0 1: There is statistically no significant effect of metacognitive regularities on work efficiency of university teachers.

H_0 1a: There is statistically no significant effect of planning on work efficiency of university teachers.

H_0 1b: There is statistically no significant effect of monitoring on work efficiency of university teachers.

$H_0 1c$: There is statistically no significant effect of evaluation on work efficiency of university teachers.

d. Significance

This research investigates the significance of work efficiency by examining the effect of metacognitive regularities on university teachers' work efficiency. The study's direct and indirect stakeholders include university teachers, students, the administration, and the university itself. These stakeholders influence one another's successes, failures, and play a crucial role in achieving the overall success of the university, which is reflected in the experiences of its teachers.

University teachers have both personal and institutional commitments that affect their practices, and responsibilities. However, these can be managed effectively through metacognitive regularities. University teachers serve as both the source and the resource for the university, and its students, who learn through teachers' reflective practices. Teachers are valuable assets to the university, but without metacognitive regularities, they face difficulties in organizing pedagogies and managing their work. This study aims to promote continuous work efficiency among university teachers through the efficient use of metacognitive regularities.

University students deserve the best professional education. As key educational stakeholders, students observe, and experience metacognitive practices in their teachers, which may encourage them to adopt these regularities in their own learning. This creates a positive feedback loop that enhances both teaching and working within the university setting. This study helps to support in improve university students' learning experiences and better academic outcomes.

The university administration was another important group of stakeholders. Similar to other non-educational organizations, the administration expects organizational achievements through the efficient application of teachers' fundamental metacognitive regularities. By promoting the importance of metacognition for work efficiency, the university administration can enhance overall university effectiveness.

The achievements of a university always provide significant benefits to its teachers. The university consistently requires highly qualified teaching staff who monitor and evaluate its progress. The university aims for social progress through the efficient work of its teachers. Encouraging metacognitive regularities is therefore a strategic approach for universities seeking to maximize their educational achievements, while enhancing overall faculty performance. University teachers and the other three stakeholders are interrelated and depend on one another. Moreover, this mix methods research study was conducted to obtain reliable results and assess university teachers' work efficiency.

e. Rationale

The concept of proficiency goes beyond simply acquiring expertise and skills; it also involves the application of established pedagogical practices and work efficiency to meet challenging demands in order to achieve it. Thenmozhi (2019) stated that metacognitive regularities were essential steps for fostering greater association with teachers and students' collective well-being in the twenty-first century across all levels of teaching and practices. Accessibility to material resources was only one aspect of well-being. Likewise, Fiona (2003) states that Cognitive awareness is practical, and the teaching profession is typically more practice-oriented than theoretical educational studies, which nevertheless offer real value. Teachers are the soul of the university, and for them, the main reason to be aware of metacognitive regularities was this very practicality. University structure has a direct impact on the learning of both staff and students, as evidenced by the work happiness of organizations and teachers (Valerie et al., 1991).

University teachers need to understand metacognitive regularities to develop new competencies and achieve future goals. Although these principles can be used to improve current work efficiency, this practice has been lacking in universities and has not been a focus of previous research. Tina and Mark (2012) stated that in educational settings, teachers, administrators, and other stakeholders often overlook the importance of mental planning for efficient work. This type of planning is crucial as it significantly impacts pedagogy and student development. The university teachers, their vital contributions to the universities achievements are inextricably linked to their personal accomplishments. This success is only possible through a perfectly balanced work schedule. Siti et al., (2012) stated a strong teacher-student relationship is essential for effective teaching and positively influences both individuals' personal development. Moreover, Sharul, Ruhanita, Amizawati, and Nor (2015) stated that The university must exercise its autonomy through effective teaching, monitoring, and evaluation.

Teachers might incorporate metacognitive regularities into their routine work and align their teaching methods with university operations for the university's betterment (Hope, 2001). These are very common discussions but remain unattempt in research studies. Jalal (2016) stated that Teacher engagement is essential for universities to adjust to change and foster innovation. This is critical because students, upon completing their education, go on to work in the nation's key industries, which must maintain a competitive advantage.

Jill, Maria, and Emmanuel (2017) stated that within the university system, teachers ignited the students' cognitive abilities and gradually encouraged them to pursue a continuous expansion of their knowledge. This process began by exploring the distinction between "efficiency" and "effectiveness." It was helpful to learn that effectiveness refers to "doing the right things at the right time," while efficiency refers to "doing things correctly." Ana and Mihaela (2018) stated that the teacher who is a manager has to be aware of the importance of proficient interactions within the literate society and interactions between university and modern society. Additionally, Sergey et al. (2019) stated that integrating metacognitive regulation into university work can be defined as achieving perfect work efficiency. Teaching and efficient work practices are made possible through metacognitive regularities. When teachers become well-versed in applying these regularities, they can guide students in a concrete and effective manner. Although teachers possess awareness, a gap often exists in their self-evaluation, highlighting the need to focus on these regulatory aspects. Focusing on metacognitive regulation has helped identify flaws, enhance enthusiasm, and develop professional competencies for effective teaching. Furthermore, Lisa and Robert (2020) stated that the effect of excellent teaching on student learning is measured by gains in student achievement, which reflect the teacher's efficacy. As a manager of human potential within the university, the teacher is dedicated to developing students' mental abilities and harnessing their cognitive skills for the benefit of the academic community. Recently, Angelina, Thomas, and Brain (2022) stated the key component is the utilization of cognitive abilities, which serves as the basis for initial work planning.

The significant insight from studying university teachers' work efficiency is that teachers must be aware of metacognitive processes relevant to the modern workplace in both private and public sector universities. Paul (2010) stated that integrating metacognitive regularities is crucial for enhancing university teachers' work efficiency, benefiting both the teachers and the university. Previous research on metacognitive regularities has been limited to non-teaching groups, creating a gap in our understanding. Therefore, further study is needed to explore how university teachers can effectively understand and apply metacognitive regularities to improve their professional performance and overall work efficiency.

Pakistani university teachers had a great necessity to apply metacognitive regularities in their work to maximize work efficiency (Organization for Economic Co-operation & Development, 2019). This study aimed to enhance teaching effectiveness by applying metacognitive regularities to organize cognitive abilities, personality, classroom management, commitment, and responsibility. University teachers are essential contributors to a sustainable and inclusive future. Although they are highly experienced in their fields, metacognitive regulation can help them become more proficient. This includes setting clearer goals for student progress and developing multiple solutions to significant challenges faced by university students.

1.2. Overview of the Book

Metacognitive regularity, recognized as a key educational priority by 2030 (OECD, 2019), involves planning, monitoring, and evaluating work processes to improve efficiency. University teachers, as central figures in higher education, manage complex academic and administrative responsibilities that require more than subject knowledge they demand systematic regulation for effective performance.

Work efficiency is essential in teaching, as it influences lesson quality, classroom engagement, and institutional success. Metacognitive regularities can significantly enhance this efficiency by supporting better organization, reflection, and adaptability. However, most existing research focuses on university teachers metacognitive practices, leaving a gap in understanding how teachers apply these regularities to improve their own work.

The problem lies in the limited exploration of how metacognitive regularities affect teachers' efficiency. This study seeks to investigate university teachers' use of metacognitive regularities, assess their work efficiency, and determine how planning, monitoring, and evaluation effect their practices.

The significance of this study extends to teachers, students, university administrations, and institutions. Teachers gain improved effectiveness, students experience better learning, administrations enhance faculty development, and universities achieve sustainability and competitiveness. For Pakistani universities, where efficiency and quality teaching are on top priority, the integration of metacognitive regularities offers a practical solution.

By bridging the gap between University teachers and their work efficiencies, this research emphasizes that adopting metacognitive regularities is not optional but essential for ensuring excellence in higher education.

Chapter Two

Literature of the Study

2.1. An Overview of the Field

"The combination of three terms, including the term itself, is known as a metacognitive regularity." Cognition, regulation, and meta-meaning are self-directed processes of the mind. Through metacognitive regulation, we can manage our thinking to enhance and accelerate learning. Furthermore, metacognition can often be divided into three components: metacognitive knowledge, metacognitive regulation, and metacognitive instruction (Mehrdad, 2016). The focus of this investigation is on metacognitive regulation. Metacognitive regulation, when combined with the work efficiency of university teachers, helps them achieve the outcomes that a particular approach is intended to accomplish. Its effectiveness depends on its range of applicability, the benefits of frequent use, and the amount of effort required to implement it (Thenmozhi, 2019). Implementing metacognitive regulation does not require expensive components or significant alterations to existing education systems. Unlike many other educational initiatives, the only significant cost associated with implementing metacognitive regulation is professional development. An effective way to gain this essential insight is by using a metacognitive tool such as thinking styles. Furthermore, they move toward becoming effective thinkers by recognizing that other people may think similarly or differently from them, and by learning to understand the implications of these thought processes and behaviors. Although several authors use different definitions, here we adopt the following one: Margaret (2022) states that the second component of metacognition in this case is metacognitive regulation. Unlike metacognitive knowledge, metacognitive regulation refers to the individual's ability to monitor and control cognitive processes. It encompasses the elements that influence cognition and the evaluative skills involved in learning or understanding. This includes the capacity to determine how, where, and why to apply a specific skill to a particular task. In this way, learners can take greater control of their own education.

1. Metacognition

Metacognition is essential for both effective instruction and self-regulated performance (Georgia & Maria, 2017). Planning what and how to teach, monitoring the progress of the lesson while teaching, making necessary adjustments, and evaluating the lesson after it concludes are all examples of the executive management of metacognition in teaching. The executive management cycle is completed by identifying ways to improve future performance in similar situations, based on both internal and external input. Therefore, within the framework of metacognition, teachers consider how their lessons will encourage and enhance students' awareness of their own reasoning as learners. Metacognition is also referred to as meta-consciousness, self-awareness, self-cognizance, and self-perception by scholars.

a. Meta-conscious

Meta-consciousness is merely a type of cognitive content; it enters consciousness in a way that is comparable to the manner in which other mental contents do (Jonathan, Michael, Benjamin, & Piotr, 2015).

b. Self-Awareness

According to Julia, Rebecca, and Jonathan (2021) that the 'self' is viewed from a societal behaviorist perspective, which holds that when the self is compared to social processes, communication and awareness are most closely related to the self-adopted position.

c. Self-Cognizance

Understanding cognitive processes is referred to as cognizance. Here, it is defined as the deliberate awareness of perception and inference in knowledge and problem-solving (Smaragda, Elena, Nikolaos, George, & Andreas, 2019).

d. Self-Perception

According to self-perception theory, people view their own behavior in the same way they observe the actions of others; therefore, an individual's behavior is influenced more by social context than by free choice (Laila & Fatima, 2020).

2. Regulatory

The second is regulatory, which is the suffix of 'metacognitive regulatory.' According to Georgia and Maria (2017) stated that Affective, intellectual, and social functioning are influenced by regulatory teaching styles. In particular, independent motivation and forms of independent regulation (such as intrinsic motivation and identified regulation) are associated with several advantageous learning outcomes. It includes effective cognitive processing, greater effort when facing challenges, more efficient time management, and focused attention. Moreover, it is a satisfying experience to be actively engaged in deeper understanding during the exercise. Several regulatory terms include managerial and governing. According to Margaret (2022) stated that Teachers are better able to meet their students' needs by providing effective solutions when they are aware of their students' metacognitive thinking. Providing support to individuals while they perform a task helps them improve their reasoning and learning strategies. They become aware of the strengths and weaknesses of their work. This awareness enables them to better organize, monitor, and evaluate their thinking the next time they engage in metacognitive activities.

a. Administrative

Despite the progress made in administration, considerable work still needs to be done. The foundations of administration rely on several valuable fields, providing strong support for upcoming agendas (Bruce, Jeremy, Janine, & Sandra, 2022).

b. Managerial

Managerial effects appear to be more significant than those of workers. Therefore, management practices are crucial for programs aimed at strengthening or developing clusters around universities (Anna, 2021).

c. Governing

The Shaoning (2018) states that the governance approach in education promotes the development of vocational institutions through effective oversight. Governance involves the shared administration of public affairs in higher education, offering a new form of representative structure.

2.2. The Background of the Field

Academic quality is largely influenced by teachers' performance. Universities develop their talented employees to enhance effectiveness and ultimately achieve world-class standards. Workplace professionalism effects both work output and an institution's ability to sustain itself. There are two categories of expertise: specific and general. The intellectual requirements necessary for individuals to perform their jobs are referred to as qualifications. Waltor (2005) stated that effective classroom management and organization by teachers lead to improved academic outcomes. Students are actively pursuing knowledge within a participatory, knowledge-centered community of inquiry, thanks to constructivist-based improvements in teaching methods. Furthermore, Siti et al. (2012) stated that effective classroom management requires teachers to maintain order in the classroom and manage student behavior.

The expertise, abilities, moral principles, and character traits that each teacher of an institution is referred to known as generic proficiency (Sharul et al., 2015). In the field of education, a resource is used efficiently when visible educational outcomes are achieved with the minimal use of resources (Jill et al., 2017). The institution's goal is to support every university teacher in becoming a contributing member of the institution and to continuously improve the delivery of education in a manner that is ethical and compliant with regulatory requirements (Ana & Mihaela, 2018). The main point could be described as 'solid, stable yet fallible, and late-forming,' while retaining a high degree of consistency among people. On the other hand, regulation can be 'highly unstable and age-independent,' changing rapidly depending on the circumstances. Teaching methods that emphasize problem-solving argue that thinking is inherently incomplete. It is not about knowing something beyond what is already known; rather, it is an ongoing activity. Knowledge is a flexible, open-ended tool for action, and the concept of applying these thoughts forms the foundation of metacognitive regulation (Thenmozhi, 2019).

The effective use of resources ensures that the wide range of educational outcomes intended by society is achieved. Di (2021) states that dedicated teachers often have strong connections to their institution, their students, or their fields of study. Declarative, administrative, and regulatory information constitute the three primary types of metacognitive knowledge, whereas active learning strategies such as planning, monitoring, and evaluating are referred to as metacognitive regulation (Angelina et al., 2022). The concept of efficiency is easy to understand: resources should be utilized as fully as possible to achieve cultural goals, especially when they are scarce which they always are. Efficiency refers to evaluating inputs in relation to the corresponding outcomes. A system that is more efficient produces greater output with a given set of input resources.

1. History

The necessity of evaluation serves as the foundation of the regulatory concept. The term 'metamemory' was originally coined by John (1970) in the early, He continued to develop the study and introduced the concept of metacognition as learners' awareness of their own cognition. He defined it as 'knowledge and understanding of cognitive phenomena.'

a. Development of Metacognitive Regularities in 1989-1990s'

Nearly forty years earlier, Ruth and Practicia (1989) stated that they introduced the idea of metacognitive regularities and the concept of the metacognitive thinker. They believed that metacognitive thinkers can detect, evaluate, and, when necessary, reconstruct existing ideas. Later, Paris, and Winograd (1990) highlighted "self-appraisal" and "self-management of cognition" as two crucial characteristics in their concept of metacognition. Although Petros (2004) stated Self-management refers to 'metacognition in action,' encompassing cognitive processes that organize or evaluate aspects of problem-solving, while self-appraisal of cognition involves observing an individual's comprehension, abilities, and psychological states during the learning process.

b. Metacognitive Regularity Progress in 1994- 2006

The conclusion indicates that if faults occur at the objective level, monitoring procedures will alert the meta-level and trigger control processes to correct the issue. This appears to be a sophisticated yet simple approach by Janete and Arthur (1994) that incorporates both metacognitive skills and knowledge. However, Katherine (1996) the majority of theory-based research has focused on meta-memory and the phenomena of judgments of learning and feelings of knowing. Later, two years after, Katherine's comprehensive theory received its initial recognition. David (1998) claimed that "Meta-affection," which emphasizes the emotional dimension of instruction, is very similar to the concept of metacognition. Petros (2004) defined that "the conscious awareness, monitoring, regulation, and evaluation of intra-personal and communal influencing activity." Next, Marcel, Bernadette, and Peter (2006) stated that an "object and level," where cognitive activity occurs at the meta-level. He regulates the object and level through general informational exchanges between the two, which convey information about the condition of the object level. The meta-level monitoring procedures and directives are conveyed to the object through control mechanisms. Continuing this, Paul (1995) defined that it is necessary for self-regulation, defined as the 'active,' goal-directed self-control of actions, motivation, and cognition by an individual for academic tasks. Then, Randy and Frank (2006) stated that Students who are adept at academic self-regulation are aware of both their own learning styles and the requirements of specific activities.

Furthermore, the two main elements are referred to as knowledge of cognition and regulation of cognition. Cognitive knowledge consists of three sub-components: Declarative knowledge, which is an understanding of one's learning style and the factors that affect performance, Procedural knowledge: It involves the understanding of techniques and methods such as interleaving, reviewing, organizing, using analogies for elaboration, and identifying key concepts and Conditional knowledge: It involves knowing when and why to employ a specific tactic, as well as planning, monitoring, and evaluating at least three fundamental aspects of cognitive regulation.

- Setting goals, applying relevant past knowledge, selecting appropriate techniques, and allocating resources are all aspects of planning.
- Self-testing activities required for control are part of monitoring.
- The monitoring of the success of these activities occurs when metacognition recurs (Daniel & Christian, 2020; Schraw et al. (2006)).

c. Brief History of Metacognitive Regularities

The speculative model of metacognition and its scope remain highly debated topics in the field of education. Planning involves selecting the steps to complete a task, predicting learning outcomes, and determining the methods to be used for discovery. As a teacher, it also includes planning the most effective techniques to achieve a specific objective.

Monitoring tasks may involve analyzing an entire work to identify any confusing passages, or solving a problem and reviewing whether the steps of the solution are correctly recognized. It can also include evaluating and improving another person's or group's productivity. Subsequently, teaching proceeds, and the teacher evaluates performance and determines a new strategy to accomplish the task. This involves continuous, on-task evaluation of performance, observation of the processes used to analyze and review information and techniques, and assessment of one's knowledge applied to complete the task (Whitebread & Pino-Pastermark, 2013). The sequence refers to the manner in which metacognitive regulatory techniques are applied throughout the course of an activity (Nilson, 2013).

According to Rachel (2014); Veenman, Van, and Afflerbach (2006) stated that the techniques most frequently addressed by industry professionals in their methods involve planning, monitoring, evaluating, and control.

d. Why Teacher Need to Use Metacognitive Regularities

Performance management has four pillars: planning for performance, increasing efficiency, evaluating progress, and achieving goals. It takes a systematic and integrated approach to help the university achieve long-term success by enhancing employee efficiency and developing the capacities of both teams and individuals. It is important to clarify the dynamics between the team and the individual. Performance management primarily focuses on achieving results. The steps taken to reach these results help both teams and individuals grow in knowledge, skills, and competencies. Recognizing, measuring, and understanding the wide variations in performance is essential (Harry, 2004). Two different types of strategies can be activated by metacognitive experience, reflecting regularities in intellect and metacognition. Although researchers often divide metacognition into three interrelated components metacognitive awareness, perception of cognitive strategies, and regulation of cognition through monitoring and control the concept of metacognition is still considered vague and ill-defined. Teaching is increasingly seen as a profession that requires critical examination of each situation, identification of priorities, establishment and monitoring of appropriate learning opportunities, analysis of their impact on students' achievement, responsiveness to students' learning needs, and personal or group reflection on the entire process (Judit, 2016). Metacognitive experience is a person's subjective assessment of the simplicity or complexity of specific cognitive functions.

i. Planning for Performance

Planning is the activity of looking ahead and developing or outlining a strategy to accomplish specific goals and objectives. It involves determining, in broad terms, what needs to be done and how to go about achieving a particular goal. Planning includes the process of making technical and logical decisions and is a methodical, deliberate, and intentional process of decision-making in advance. It is an integral aspect of every individual's social efforts (Charles, 2018). Performance management in universities is a modern approach to evaluating the quality of education they provide, as well as a fusion of contemporary methods for managing human resources in education. The assessment of a university's work will be continuously refined and improved through planning within the process of ongoing inquiry and study (Qiao, 2019).

ii. Increasing Efficiency

The way efficiency is applied in professional capital for instructional purposes consists of three components: human capital, which is an investment in a teacher's knowledge and abilities; social capital, which is an investment in the relationships between teachers and the nature and frequency of interactions that enable them to work effectively as a team; and decisional capital, which is the understanding and expertise required to make accurate evaluations of students, developed over a long period. In team environments, increased expertise may enhance not only an individual's performance but also that of less experienced colleagues who gain insights from them (Anne et al., 2019). Moreover, effective instructors can have a lasting, significant, and positive impact on their students' lives. Teaching is a lifelong profession (Weilin, 2023).

iii. Evaluating Progress

Supporting students in achieving a set of focused learning goals is the primary objective of education and classroom teaching. These educational goals or outcomes encompass changes in the mind, heart, and body. Monitoring is necessary to determine whether students have undergone the intended changes. In this context, assessments and additional indicators are used to continuously evaluate learners' development. Similarly, supervisors and teachers assess students' academic progress and effectiveness annually at various levels (Nelson, 2018).

iv. Goal Achievement

Teachers who pursue mastery goals are primarily focused on the educational process, as well as the enhancement and development of their skills. In contrast, individuals who aim for performance goals are mostly self-focused, goal-oriented, and concerned with how they compare to others. Performance is defined based on acceptable societal judgments. The standards and observable aspects are experimentally distinct and differentially influence educational achievement (Martin, Oliver, & Markus, 2018). Moreover, Stanislav, Donatas, Nele, and Tomas (2025) states that metacognitive abilities are necessary for effective reflection on experiences and are considered essential for the development of responsive competence.

e. Self-Monitoring Used for Higher-Order Metacognitive Regularities

People may select, evaluate, and adjust cognitive patterns with the aid of metacognitive knowledge, which is crucial for creative thinking. Empirical studies have shown that individuals' metacognitive knowledge supports knowledge innovation. Although decision-making is a fundamental teaching skill and an essential aspect of an instructor's professional career, it requires careful application (Tina & Mark, 2012).

The higher-level cognitive cue of processing fluency can be used to indicate metacognitive experience. Xiaoyu, weijian, and Liren (2019) stated that by using metacognition to monitor other processes, the creative thought process essentially "self-monitors" through the interaction of systematic review and elemental information. Classroom learning is significantly influenced by planning, whether mental, written, or a combination of both. Preparation and reflection help teachers become more skilled at managing the various aspects of teaching, as many decisions in the classroom must be made without sufficient time for contemplation or planning. In the classroom, decision-making is guided by evaluation. Decisions made during instruction differ from those made during planning, as they are made quickly and without the opportunity to reflect or consult new information.

f. Teacher Work Efficiency

The awareness of one's own reasoning and metacognitive evaluation is important. University teachers develop a cognitive framework through the memory of concepts. The deliberate integration of this knowledge into team processes and personnel management tasks is highly effective (Fiona, 2003). Gloria (2005) stated that although participants enjoy being exposed to new teaching concepts and methods, the curriculum does not tailor or provide a supportive toolkit for college instructors. A strong connection is needed between the student, the teacher, and the knowledge being learned. Metacognitive evaluation involves using systematic techniques in problem-solving, as well as organizing and generating meaningful information about problems. Teachers' work lives and hours are influenced by their roles and responsibilities (Siti et al., 2012). University teachers are both instructors and managers. As a result, they understand the academic components that aid in decision-making and in adjusting flexible elements according to the situation. Workplace efficiency must be enhanced through monitoring a wide range of cognitive activities. This includes the actions and relationships within categories of concepts, metacognitive knowledge, metacognitive situations, goals, and strategies that support and facilitate effective performance.

g. The Concept of Work Efficiency

Teaching is not solely a technical or rational profession. Defining standards of professional performance is highly challenging. It is believed that teaching is 50% science and 50% the craft of education (Elena, 2014). A teacher's productivity is evaluated by comparing it to the standards set for teachers performing similar tasks. The length of tenure influences the intellectual capacity and efficiency of the teacher. This suggests that a teacher's productivity and performance contribute to the overall development of an institution.

h. The Importance of Teacher Work Efficiency

The most essential component of success and responsibility is work efficiency, as it determines the potential for future tasks. It evaluates one's current level of knowledge while performing a cognitive task. It is crucial to keep a record of personal experiences and failures and to update metacognitive knowledge regarding the effectiveness of one's work (Roger & Vincent, 2013). Additionally, Khoirotun, Aslamiyah, Noorhafizah, and Novitawati (2025) stated that teachers play a significant role in maintaining academic standards because they apply their teaching skills, and instructors must be effective in their roles.

2.3. Key Concepts

1. Theoretical Base of the Study

A study's theoretical base is the collection of previous theories and concepts that serve as the foundation for and direction of the investigation. The theoretical base of metacognitive regularities in the provided report revolves around a framework that divides metacognition into two main components: metacognitive knowledge and metacognitive regulation. This structure is primarily informed by the work of Schraw and Moshman (1995) and Nelson and Narens (1990) (as cited in Julie et al., 2021) and is used throughout the report to investigate metacognitive regularities for university teachers. This relates, teachers regulates their thinking during work. It comprises: Planning, deciding what regularities to use. Monitoring, assessing one's understanding and the effectiveness of work efficiency. Evaluation, reviewing and judging the effectiveness of regularities. These regulatory processes enable teachers to adjust their work based on evaluation, a key marker of expert-like thinking.

Harris and Rutledge (2007)(as cited in Siti et al., 2012) Stated work efficiency model by visualize teacher quality based on their practices and Brown (2004) emphasized that effective classroom management includes the teachers' ability to respond appropriately to the emotional, social, cultural, and cognitive needs of the students. The theoretical base of the Teacher Work Efficiency model determine four dimensions of teachers. It reflects a complete view that quality teaching practices result from mental skills (cognition), personal traits (personality), and the ability to manage and engage students in the classroom. Therefore, cognitive and personal attributes must complement each other for comprehensive teacher effectiveness. The model offers a robust structure for evaluating teacher work efficiency beyond mere student outcomes, making it a valuable tool for universities. Commitment and responsibilities refer to a teacher's dedication to their professional duties and managed classroom obligations.

The level of commitment also affects teachers' quality of work efficiency. According to Piaget (1936) theory of cognitive development, children fundamentally go through four different phases of cognitive development, such as; Sensorimotor, Pre operational, concrete operational and formal operational. Every stage has certain turning points at which an individual starts to understand a more complex awareness of their surroundings. In the meantime, another psychologist shared his opinions about the theory of cognitive growth was Vygotsky (1934) states the sociocultural theory of development, a child's capacity to learn through social interactions and their culture is facilitated by the social process of student learning. Piaget's theory of cognitive development differs greatly from Vygotsky's theory. The both concept motivate to develop current study. People who regularly employ metacognitive strategies are guaranteed to be able to distribute cognitive resources as efficiently as possible, particularly in situations with high demands. Through the optimization of metacognitive regularities efficiency improves in the workplace or in the classroom.

Lastly, this study assessed the effect of metacognitive regularities on work efficiency of university teachers through the above-discussed theoretical base. Metacognitive regularities are fundamental at every stage, encouraging efficient working or task completion. Initially, Flavell's (1971) created the theory and came up with the name metacognition. Knowledge and cognition about cognitive phenomena, as he put it is the term he used to describe a learner's awareness and comprehension of their own cognitive processes. Knowledge how people think about their own thinking is a key idea for learning and cognitive growth and Flavell's study established the groundwork for this knowledge. Moreover, Sweller (1988) Cognitive load theory, he stated that as working memory has limited capabilities, regularities that are effective in reducing unnecessary cognitive burden can improve work efficiency.

The three characteristics of regulation learning are their interconnected motivational processes, teachers' employment of regulation teaching practices and their response to self-oriented feedback regarding the work efficacy of their University working. Teachers who engage in metacognitive practices can improve their work efficiency. The relevance of metacognitive regularities to work efficiency, particularly in the context of teaching practices, can be understood through the lens of continuous improvement. Furthermore, the study provided a rich description of metacognitive regularity, focused on the work efficiency of university teachers.

a. Conceptual Framework of the Study

This conceptual framework of the study was established to analyze the practices of teacher work efficiency. It can help in tracking metacognitive regularities teachers' practices. It can give progress in promoting effective university work and pedagogies. Separate investigations have been conducted on these two variables. The effect of metacognitive regularities on work efficiency has been studied with other subjects, but none of the research has studied the University teachers. The results of the present research study explore the effect of metacognitive regularities on university teachers' work efficiency in their practices, was adds to the text of knowledge. The conceptual framework of the study consisted of models, i.e. metacognitive regularities (Julie et al., 2021) and work efficiency (Siti et al., 2012).

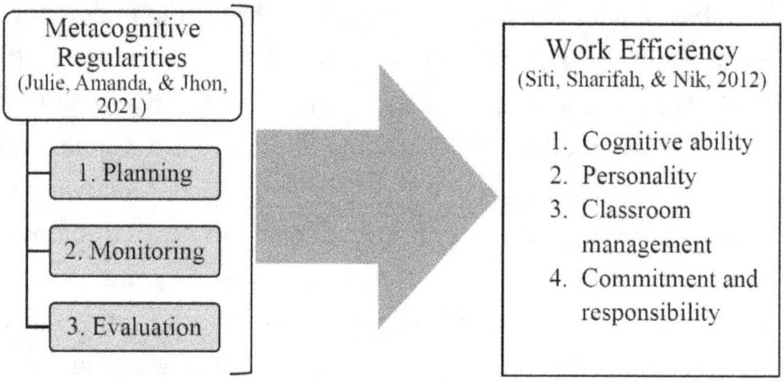

Figure No.1.1. Conceptual Framework of the Study

i. Metacognitive Regularities Model

Metacognitive regulation refers to learners' capacity to monitor and evaluate their knowledge and vice versa metacognitive knowledge is learners' knowledge or views about the elements that affect cognitive skills. It includes decision making ability to decide whether, when and where they can apply specific expertise for a specific reason. Teachers can take responsibility for their work in this context. (Julie et al., 2021). Metacognition depends on the consciousness of efficiency. The choice of appropriate planning and the controlling of resources that directly affect university teachers' practices.

1) Planning

Planning refers to a management procedure that is based on input from internal as well as external sources. It determines effectivity in every situation to move forward. (Julie et al., 2021).

2) Monitoring

The monitoring is the follow-up of abilities and techniques for the potential result. It can be learned by applying previous observations' results in new situations (Julie et al., 2021).

3) Evaluation

The method of evaluation usually involves connections between previous knowledge and new information. Organizing and structuring content, reorganizing one's cognitive processes, using schemes and focusing on proof and reasoning, as well as making connections between self-experiences and realistic scenarios (Julie et al., 2021).

The procedure for identifying how teachers always reflect on their own intellectual and teaching practices is called a metacognitive regularities indicator. It benefits in recognizing changes in the way a teacher plans, monitors, and evaluates their practices over time.

ii. Work Efficiency Model

Teacher effectiveness is a reflection of an individual's professional qualities, including expert practice, effective classroom management, student participation, and ongoing improvement. It is influenced by best practices, excellent teaching techniques, dedicated work and proper involvement in-class sessions (Siti et al., 2012). Cognitive skills are essential for comprehending data, remembering team objectives during work, and concentrating while performing tasks in an organization.

1) Cognitive Ability

The intellect or cognitive ability is a broad mental capability encompassing logic, problem-solving, creative thinking, understanding complex concepts, quick learning, and insight (Siti et al., 2012).

2) Personality

Personality is the specific feature of humans that relates behaviors and illustrates traits of a competitive attitude in comparison to moderate individuals (Siti et al., 2012).

3) Classroom Management

Classroom management involves teachers' methods to maintain order in their classes and considers group dynamics and circumstances because management is a characteristic of sociological systems (Siti et al., 2012).

4) Commitment and Responsibility

Academic standing involves intellectual commitment to group goals, honesty in scholarly work, and responsibility for becoming knowledgeable about academic practices (Siti et al., 2012).

A work efficiency indicator for teaching is a criteria or set of criteria used to evaluate how effectively a teacher performs their work, balancing the quality of practices and outcomes. These indicators help school administrators, students, the university teachers themselves to identify strengths and areas for improvement.

2.4. Other Supporting Theories and Models

Several frameworks important for metacognitive processes have been introduced by various authors, each with its own distinct characteristics. These frameworks are valuable for pedagogy and for enhancing teachers' effectiveness. Details of the mentioned models and theories are provided below.

1. Metacognition & Social Metacognition Theory

Social metacognition enhances cognitive abilities, promotes social guidance, and improves interpersonal interactions by fostering shared thinking, task sharing, and the visibility of each other's metacognitive processes (Ming and Sze, 2009). They worked on the given theory of metacognition, which was described as individual, and social metacognition was analyzed for the group using the same dimension. Social metacognition is an extension of metacognition into group interactions. Similar to individual metacognition, social metacognition involves group members monitoring and controlling one another's knowledge, emotions, and actions. The benefits of metacognition and social metacognition were examined through dimensions of difficulties, implications for student learning, and approaches for teaching metacognition in schools.

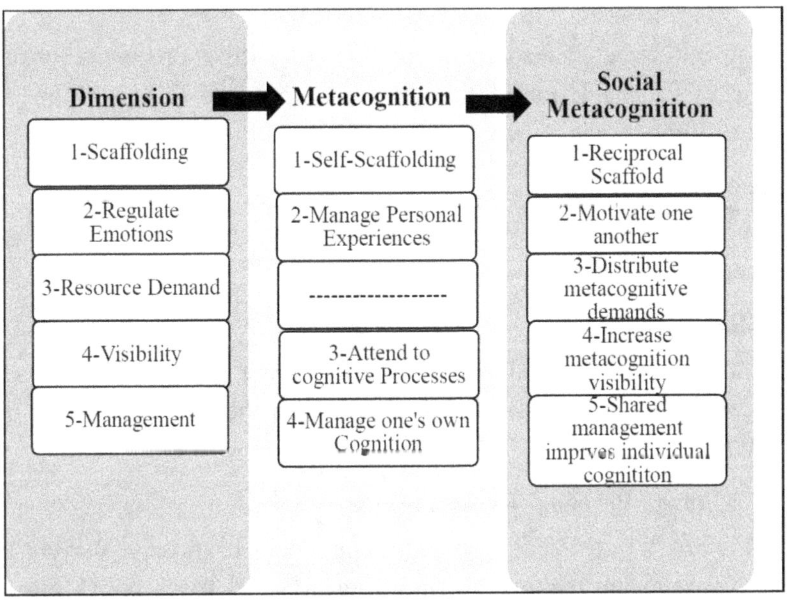

Figure No. 2.1. Benefits of Metacognition & Social Metacognition Theory (Ming & Sze, 2009)

a. Dimension

The dimension emphasizes the interplay between individual and group dynamics in cognitive regulation (Ming and Sze, 2009).

b. Scaffolding

Scaffolding involves providing temporary support to individuals or group to enhance metacognitive skills (Ming and Sze, 2009).

i. Regulate Emotions

Emotions play a crucial role in metacognitive processes by influencing motivation and cognitive control (Ming and Sze, 2009).

ii. Resource Demand

It refers to the cognitive and emotional resources required to engage in metacognitive processes (Ming and Sze, 2009).

iii. Visibility

It is often less visible since it occurs internally and more visible due to group interactions (Ming and Sze, 2009).

iv. Management

The effective management of metacognitive processes ensures successful regulation of thoughts, emotions and actions (Ming and Sze, 2009).

c. Metacognition

The metacognition refers to the awareness and regulation of ones' own though processes (Ming and Sze, 2009).

i. Self- Scaffolding

This is focusses on how individuals create internal structures by actively organizing and guiding their own thinking (Ming and Sze, 2009).

ii. Manage Personal Experiences

This involves reflecting on past experiences to improve current and future decision making by analyzing personal successes (Ming and Sze, 2009).

iii. Attend to Cognitive Processes

This involves actively monitoring and evaluating cognitive activities to ensure that cognitive resources are allocated efficiently and effectively for the task at hand (Ming and Sze, 2009).

iv. Manage One's Own Cognition

It is regulating thinking processes to underscores the importance of being proactive in controlling one's mental activities to achieve desired outcomes (Ming and Sze, 2009).

d. Social Metacognition

It is refers to the shared process of thinking about one's thinking in a social context to focuses on collaborative environments to enhances individual cognition through group interactions (Ming and Sze, 2009).

i. Reciprocal Scaffold

Being a member of a group support and build each other understanding (Ming and Sze, 2009).

ii. Motivate One Another

It is social interaction which fosters motivation by promoting accountability and shared goals (Ming and Sze, 2009).

iii. Distribute Metacognition Demands

In group settings, Individuals share the cognitive load by divided complex task and encourages efficient use of resources (Ming and Sze, 2009).

iv. Increase Metacognition Visibility

This visibility enables group members to learn from each other's approaches (Ming and Sze, 2009).

v. Shared Management Improves Individual Cognition

A collective focus on metacognition ultimately improves personal cognitive abilities (Ming and Sze, 2009).

The theory concludes that metacognition can help students learn and solve problems by promoting self-scaffolding and managing personal experiences. Similarly, social metacognition distributes metacognitive demands among group members, increases the visibility of one another's metacognition, and improves individual cognition, resulting in reciprocal scaffolding and greater motivation.

2. Metacognition Framework Theory

According to Rachel (2014) stated in his research work that metacognition has two domains knowledge and regulation the previous idea of metacognition was divided into two frameworks based on the earlier models. The process a learner goes through to implement and maintain cognitive functioning, behaviors, and metacognitive processing to achieve a specific goal or set of goals involves knowledge, which is an essential element.

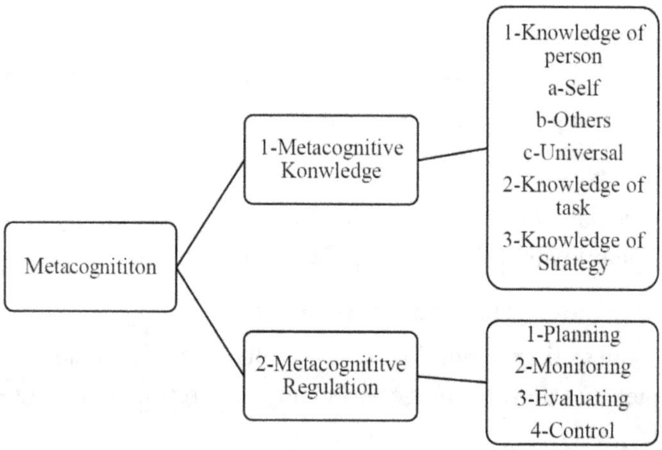

Figure No. 2.2. Metacognition Framework Theory (Rachel, 2014)

a. Metacognition

The metacognition as a structured way of understanding how individuals think about and regulate their thinking processes (Rachel, 2014).

i. Metacognitive Knowledge

It refers to the awareness and understanding of one's own cognitive processes (Rachel, 2014).

1) Knowledge of person:

This refers to an individual's awareness of their own cognitive abilities and limitations.

2) Knowledge of task:

This involves understanding the nature of the task at hand and the cognitive demands it places.

3) Knowledge of strategy:

This pertains to knowing which strategies to apply to achieve goals effectively.

ii. Metacognitive Regulation

It refers to the active processes individuals use to control and direct their cognitive activities (Rachel, 2014).

1) Planning: Setting goals and selecting strategies for approaching tasks.

2) Monitoring: Checking progress and assessing the effectiveness of strategies during the task.

3) Evaluating: Reflecting on the outcomes and determining whether goals were achieved.

4) Control: It is the decision making mechanism that enables individuals to adapt their behavior, allocate resources and employ strategies to optimize task performance.

This theory aimed to describe the metacognition of engineering students in the naturalistic setting of study groups and the contextual factors that supported this engagement. Analysis of the observational field notes found that undergraduate engineering students engage in metacognitive engagement through metacognitive knowledge and metacognitive regulation at a higher rate.

3. Model of Metacognition

The second model presented Markeya (2016). In this, he explains why it is crucial for teachers to actively encourage students' metacognition. The brief description of "thinking about thinking" or "cognition about cognition" is frequently used to describe metacognition. She created a conceptual framework for a commonly occurring component of metacognition after analyzing work by Flawell et al., 1979-2011. It is more accurate to say that it also includes understanding, consciousness, and control of one's own cognition as well as cognitive processes in general.

Two main elements of metacognition are the presentation and description of the regulation of cognition, the situations of learning. In addition, he asserts that training at all levels of institutions can improve metacognition, which develops as a component of cognitive development. He creates metacognition conceptual framework composed of the knowledge of cognition and regulation of cognition as its two fundamental categories. Declarative, procedural, and conditional knowledge of cognition comprises three categories while planning, monitoring, controlling, and evaluating knowledge of cognition contains four more. These subcategories specifics are listed below.

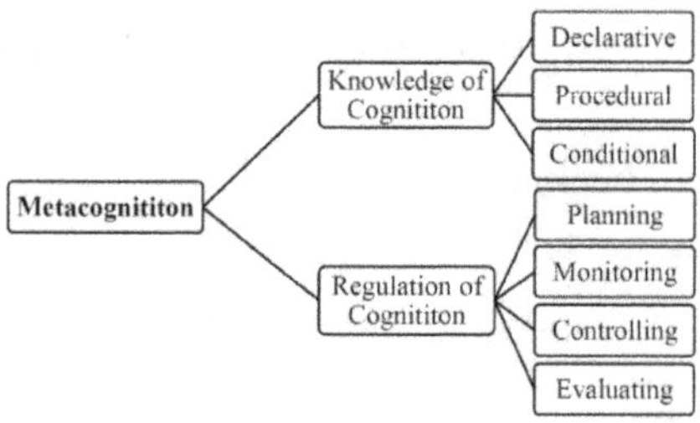

Figure No. 2.3. Model of Metacognition (Markeya, 2016)

a. Knowledge of Cognition

A person's knowledge of cognition involves his or her understanding of tactics, his or her own ways of thinking, and humans overall as cognitive beings (Markeya, 2016).

i. Declarative

Declarative knowledge comprises information regarding one's own cognitive capabilities and variables affecting one's demonstration and education.

iii. Procedural

University scholars with procedural expertise can take notes during lectures and write a successful introduction prior to drafting an assignment.

iv. Conditional

Understanding when and why to utilize a specific approach is referred to as conditional knowledge.

b. Regulation of Cognition

The "proactive" component of metacognition is the regulation of cognition. This category of abilities encompasses cognitive planning, monitoring, control, and evaluation processes (Markeya, 2016).

i. Planning

Establishing targets, choosing solutions in advance, and deciding the sequence of the tasks that will be done are all examples of planning cognition.

ii. Monitoring

Monitoring cognition means being conscious of one's understanding mental processes, and tactics used to complete a task.

iii. Controlling

Administration of memory, blocking undesirable responses, and limiting thoughts are some examples of cognitive control approaches.

iv. Evaluation

Identifying and fixing mistakes, comparing results to objectives, reporting on performance, and assessing the effectiveness of one's learning are all parts of evaluating cognition.

c. Suggestions for Promoting Metacognition Instruction

After doing the study, Markeya (2016) discovered that adding extended metacognitive training to a college course caused students' metacognition to rise risings that weren't observed in students attending a course similar to this one without the metacognitive training. Furthermore, he addresses the metacognitive teaching of teachers, which was also necessary to foster metacognition in students, in their systematic review of their work. He states metacognition is linked to a variety of achievement outcomes and those who are building their mental ability will benefit most from metacognitive training. It is divided into two categories: implicit instruction and explicit instruction. One of the teaching methods may be used by the teacher; however, neither may be used together. Implicit instruction happens when the nature of a lesson or associated activities contributes to the likelihood that students engage in metacognition without explicitly addressing the "how" or "why" of doing so. The teacher will typically do this by stating, describing, or explaining the advantages of metacognition. Such as instruction about using "how" or "why" called metacognition explicitly.

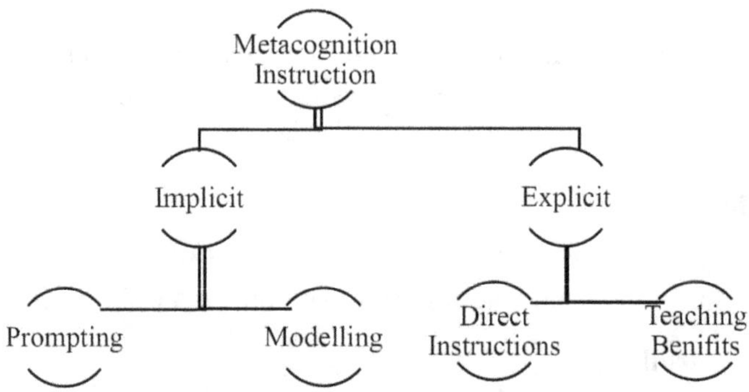

Figure No. 2.4. Suggestions for Promoting Metacognition Instruction (Markeya, 2016)

d. Metacognition Instruction

It is Emphasize strategies to help students develop their ability to think about their own thinking (Markeya, 2016).

i. Implicit

If is refers to delicately encouraging students to develop their metacognitive skills.

1) Prompting: the teacher embeds questions and tasks that prompt students to think about their thinking processes.

2) Modelling: It refers to the teacher demonstrates metacognitive strategies during their teaching process.

ii. Explicit

It refers to teaching students about metacognition to clearly explaining applications.

1) Direct instruction: It refers to explicitly teaching students the skills and strategies associated with metacognition.

2) Teaching benefits: It is Enhancing retention and developing skills that will serve students beyond the classroom.

It is believed that both explicit and implicit cognitive processing education are significant. Strategic instruction is the term used to describe metacognition teaching, which frequently entails teaching pupil's strategies. Regularities are actions that help speed up learning or task completion. The five steps of Embedding Metacognition Teaching and three basic guidelines for conducting efficient metacognition training give teachers a placc to start when adding more metacognition coaching to their lessons.

4. Promoting Socially Shared Metacognition Regulation

The Dongho and Cheolil (2017) states that participants' interactions with regard to collective preparation and understanding development were positively impacted by the collaboration programs applied to enhance socially shared metacognitive control. Furthermore, ability and reflection on job performance are covered by metacognition regulation.

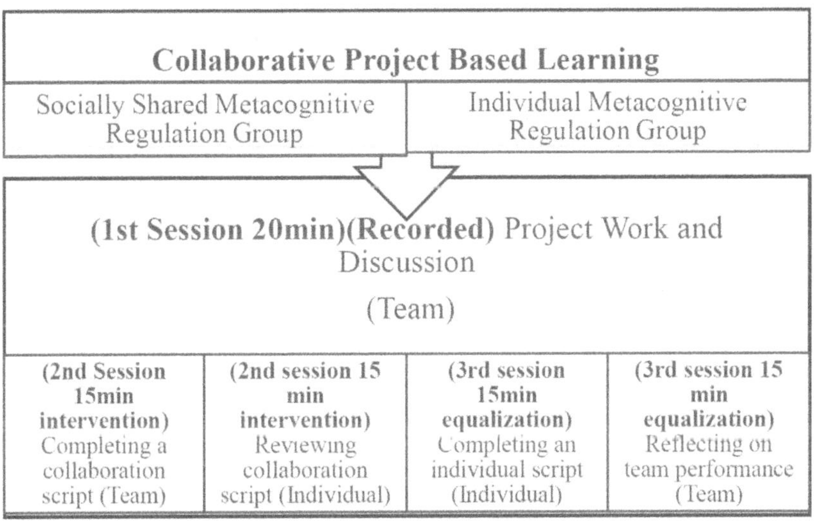

Figure No. 2.5. Promoting Socially Shared Metacognition Regulation (Dongho & Cheolil, 2017)

a. Socially Shared Metacognitive Regulation Group

It refers to highlights collaboration can enhance the overall cognitive and metacognitive development of a group and more adaptive approaches to complex tasks (Dongho and Cheolil, 2017).

b. Individually Metacognitive Regulation Group

It refers to the processes where an individual reflects on and regulates their own thinking and learning (Dongho and Cheolil, 2017).

Collaborative project-based learning is a pedagogical process that aims to facilitate students' investigation and knowledge gain for intellectual goals. Although the empirical study revealed the process of team knowledge construction. The results showed that an important difference existed between the two conditions in terms of trends in intellectual and planning interactions. The socially shared metacognitive regulation teams revealed an increasing engagement with the project content. Their collaboration with the content implied that the socially shared metacognitive regulation teams made a successful shift from managerial connections to intellectual interactions toward knowledge construction. The existence of intellectual dealings is a key indicator of a high-level, socially shared metacognitive regulation process. Unlike low-level, socially shared metacognitive regulation confined to exploring task demands. Our finding from the interaction analysis indicated that the individual metacognitive regulation teams exhibited more confusion in amending lists and allocating tasks, which continued till the last discussion. In the last parts of the project required substantial commitment to constructing a deeper understanding of content, they should have prepared for collaboration at an initial phase.

5. Metacognitive Self

Hanna, Roman, Pawel, and Rex (2019) state that Meta-cognitive self performs crucial self-regulatory tasks. It appears that researchers may now effectively evaluate each person with respect to metacognitive self. Despite the writers mentioned above and others, who came before them describing that control affects several other cognitive processes, there is currently an interest in an era in doing additional research on this area of study.

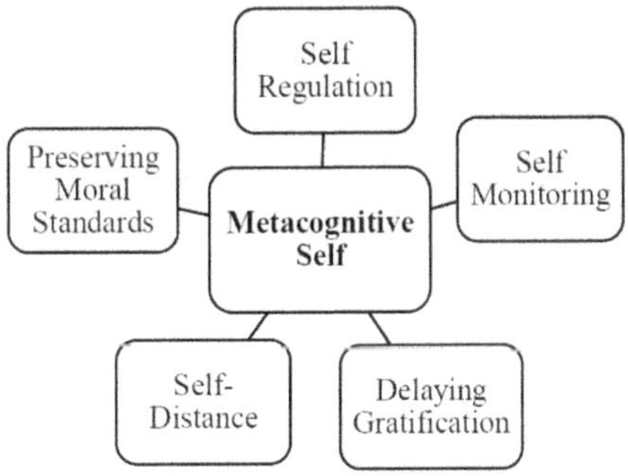

Figure No. 2.6. Metacognitive Self Theory (Hanna, Roman, Pawel, & Rex, 2019)

a. Metacognitive- Self

It refers to individuals can enhance their capacity for learning personal development and overall functioning (Hanna et al., 2019).

i. Self-Regulation

Adjusting ones cognitive strategies, behaviors and emotional responses in response to challenges or changing contexts (Hanna, et al., 2019).

ii. Self-Monitoring

It refers to the process by which individuals actively observe and regulate their thoughts, emotions and behaviors (Hanna et al., 2019).

iii. Delaying Gratification

Delaying gratification refers to the ability to resist the temptation of immediate rewards in favor of achieving larger, more valuable and meaningful outcomes in the future (Hanna et al., 2019).

1) Self-Distance

Self-distance is a vital metacognition skill that contributes to self-regulation and adaptive coping align with the broader goals of metacognitive self-theory (Hanna et al., 2019).

2) Preserving Moral Standards

This ensures individuals maintain their moral integrity, negative ethical complexities and respond adaptively to moral challenges (Hanna at al., 2019).

The metacognitive self-theory in this framework used the components; including self-regulation, self-monitoring, delaying gratification, self-distance and preserving moral standards. Every components collectively judge the personality of an individual metacognitions.

6. Temporal Regularity

Additionally, competent performance is evaluated based on proficiency, which is operationally referred to as a steady rhythm or time-based regularity rather than speediness (Lisa and Richard, 2020).

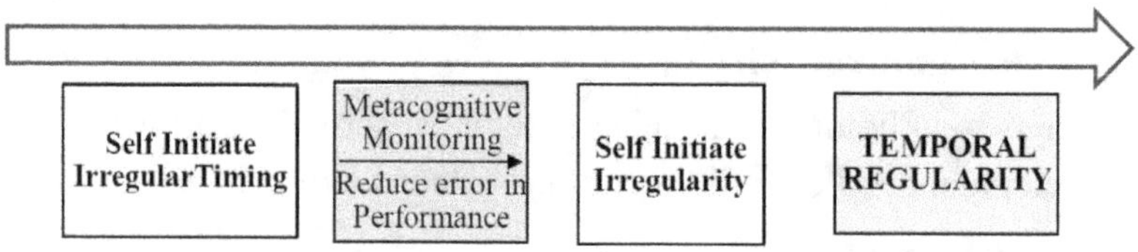

Figure No. 2.7. Temporal Regularity (Lisa & Richard, 2020)

a. Self-Initiate Irregular Timing

It begin to engage in an action or activity but the intervals between these actions are irregular and not predefined (Lisa and Richard, 2020).

i. Metacognitive Monitoring:

It is refers to the process by which individuals track assess and reflect on their own cognitive processes over time.

ii. Reduce Error in Performance:

Individuals or system can careful review their performance, avoid repetitive mistakes and optimize decision making processes over time.

b. Self-Initiate Irregularity

It refers to the Phenomenon where irregular patterns or behavior emerge independently with in a given system without external influence (Lisa and Richard, 2020).

c. Temporal Regularity

It is refers to the predictable patterns of events or stimuli over time (Lisa and Richard, 2020).

The concept examined performance as monitored based on fluency, which is operationally defined as temporal regularity rather than speed. Since error is regularly associated with flexible timing, it tested the possibility that people use various timing as a metacognitive cue. Using a sequential counting task, which may be representative of the broader class of skilled and multi-step tasks, it found that flowing between irregular and regular timing led to greater assurance ratings when the timing associated with the task was regular. The regular, consistent timing, when compared directly to irregular timing, produced feelings of fluent task performance, leading to increased confidence that found a split between perfection and belief, support the argument that individuals relied on monitoring of fluency to support their metacognitive decisions.

7. Model of Metacognitive Process by Damien, Bert, and Wouter (2021)

Damien, Bert, and Wouter (2021) define meta-knowledge as an upward progression from object-level to meta-level. The downward progression from the meta-level to the object-level is referred to as meta-control. Therefore, starting from the bottom-up monitoring and top-down management of object-level processes is the way metacognition is understood. Metacognition has mostly been studied by educational academics using the Self-Regulated Learning theory as a framework. Setting objectives, planning, organizing, self-monitoring, and self-evaluation "at multiple times during the training" are all examples of metacognition.

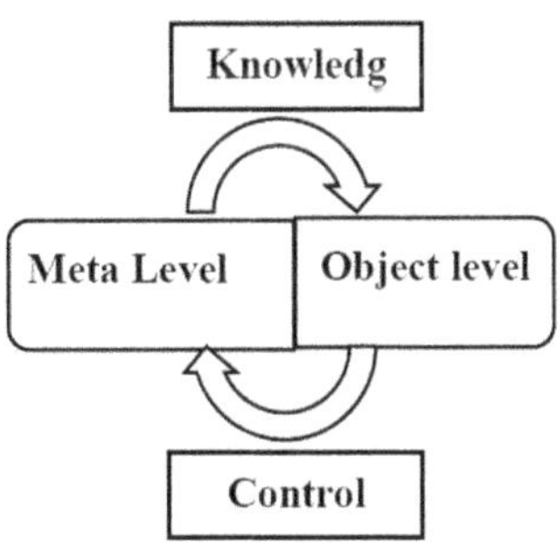

Figure No. 2.8. Model of Meta Cognitive Process (Damien, Bert, & Wouter, 2021)

a. Meta Level

A contrast amongst levels of conceptualization is called a Meta level. General concepts, "arguments about arguments," or "thinking about thinking" are the topics covered.

b. Meta Objective

The object level typically focuses on the current problem at hand. The method of interpreting imagination by making use of people or circumstances from the outside world.

c. Control

A command over concerning a control, for instance, a control that describes how to handle the required information want or existing status of information for another control.

d. Knowledge

Meta-knowledge is knowledge regarding information. Meta-knowledge also refers to understanding what other people know, what information other people require, and the way one's knowledge can be used appropriately. It includes knowledge required understanding, its source, practicality, and trustworthiness.

There is not a clear description of the difference between meta-knowledge and meta-control, although the term "self-oriented feedback loop" is frequently used to describe the interaction between reflective and regulatory processes. As protocols operationalize self-awareness and strategy knowledge, they also operationalize meta-control when they operationalize the choice, application, and planning of learning strategies in educational sciences.

8. Metacognitive Regularities with Three Layers

Metacognition was first described (Flavell, 1979) further develop the concept were Julie et al (2021). This proceeds concentration on the methods users choose, the factors that influence those regularities and the solution to those issues through the application of metacognitive regulation. It improves productivity in the educational sector, such as when taking classes, and tests or working as a team member with institutions. Social cognition is often referred to as socially shared metacognition. It is a relatively recent topic of research. The science of metacognition that is covered by working with teachers and learning organization.

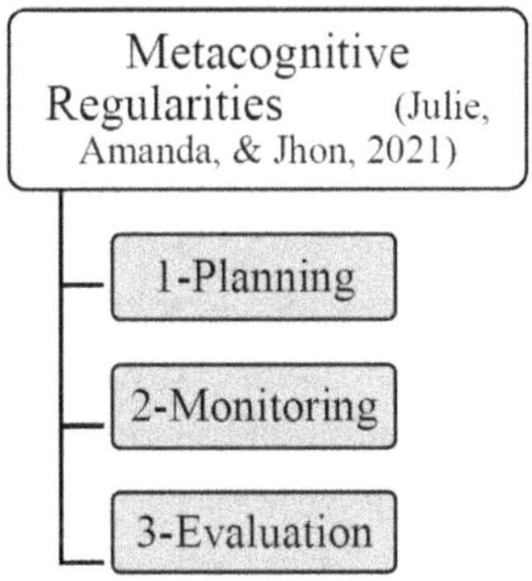

Figure No. 2.9 Metacognitive Regularities with Three Layers (Julie etal., 2021)

This concept offers an impressive categorization based on an accurate structure. A substantial amount of cognitive regularity usage for work effectiveness is also explained by this concept. Additionally, this theory has value for lifelong applications that affect pedagogical and organizational effectiveness. The specifics of all the layers of this framework are listed below:

a. Planning

The planning is concerned with activating metacognitive understanding, which comprises expertise in cognitive tasks and cognitive techniques. Emmy, Sjef, and Theo (2017) stated that time management is a crucial part of planning, as creating calendars for teaching and allotting plans for various tasks. Then Indra (2023) states the process of selecting one of the multiple possibilities for the objectives that an organization wants to attain through a sequence of interconnected actions is known as planning. All are to be prepared through educational scheduling.

b. Monitoring

Monitoring is an entire structure that carries out numerous tasks. The monitoring's repeatedly gathers of information. The analytical nature set it from related instructional and mental operations (Gylmira, Altynay, Sholpan, Aitzhanova, & Marat, 2016).

c. Evaluation

The evaluation analyzes a wide range of specific abilities and behaviors, including preparation, managing the classroom, directions, and material knowledge. Eric and John (2012) stated that a scoring rubric that describes the way every skill and practice is performed is used by evaluation professionals. Teachers are evaluated in terms of how to support, collaborate, improve teaching skills and engage in leading professional development. Consequently teachers provide collegiate evaluation for the purpose of building professional capacity and practice improvement. Moreover, John and Mary (2019) stated that as a leaders, faculty and staff create and maintain an atmosphere of dedication to a common vision to set the targets and aims.

9. The 9-Layered Model of Meta-Learning and Meta-Intelligence

In this model Athanasios, Eleni, and Charalabos (2023) state that through the use of metacognition. The current model seeks to explore the idea of meta-learning and outline the meta-levels of learning. Additionally, smart technologies can serve as a favorable environment. The meta-learning is investigated for training methodologies. Innovative tools are prepared to be used, modifying the environment of learning, offering chances for revolutionary experiences in education and promising more awareness, self-directed and self-motivated learning. The term "meta-learning" describes a collection of mental meta-processes used by learners to intentionally develop and maintain their own particular learning models. It ensure the transfer to the greatest level. Meta-learning entails a group of meta-skills to be gradually and systematically modified.

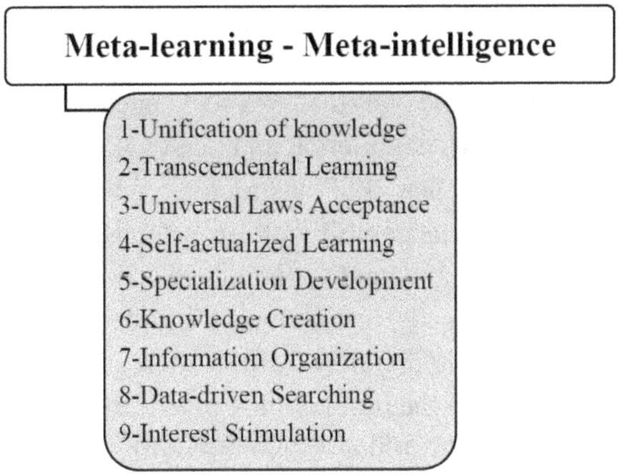

Figure No. 2.10. The 9-Layered Model of Meta-Learning – Meta-Intelligence (Athanasios, Eleni, & Charalabos, 2023)

a. Ideas of Metacognition by Athanasios, Eleni, and Charalabos (2023)

The concept is that driving change in education will unavoidably result from events that are about to change modern society forever. In order to receive, decode, retain, and retrieve information, teacher engage in a variety of internal brain processes known as cognitive learning. Looking for meaning and comprehension is how constructivism views learning. Therefore, active and self-directed learning is necessary for people to learn more effectively. According to recent research, regular or unpredictable brain activity holds a great deal of promise for providing fresh perspectives on self-directed learning (Athanasios et al., 2023). After investigation, they found the meta-level of the educational hierarchy's issue being related to the conversion of data into information that is both meaningful and useful. The top meta-level, learners will use a variety of meta-strategic manipulations to organize, build, and extract meaning from information in the form of knowledge. Athanasios et al. (2023) realized that in order to make education greater independent, autonomous, and self-regulated. It is necessary to have skilled trainers and teachers who perform at a high level in a particular field of knowledge. Purifying abilities explain the meta-capacity to be selective and discover the variables that ensure success. In scientific study, for example, it is related to the ability to recognize the boundaries of knowledge. Facilitator as a mentor, helps Meta-learners distinguish between cognitive and emotional environments. Select the ones that are most beneficial through support of their goals, success, completion and personal growth. Knowledge-restructuring approaches and instructor control over discipline networks can both considerably help in establishing areas of study. The start of the stage for fundamental sustainability author develops an eight-layer model based on the regulatory, meta-abilities and meta-skills that students deliberately utilize to regulate their cognitive processes and produce the best learning outcomes. It encompasses the capacity for learners to strategically employ their mental faculties to accomplish higher goals. The meta-ability to monitor, regulate, and adjust cognitive functions for individuals' awareness that make concern in their capabilities and approaches.

b. Meta Learning

The student can advance beyond their current understanding and reach the next stage of comprehension, they are said to be meta-learners (Athanasios et al., 2023).

i. Unification of Knowledge

A logical theoretical framework of science known as a theory of everything, the final theory, the ultimate theory, unified field theory, or master theory explains and connects every aspect of the universe.

ii. Transcendental Learning

As soon as the students are able to progress beyond their current understanding and reach the next phase of understanding, they are considered to be a meta-learners or more specifically meta-thinkers.

iii. Universal Laws Acceptance

We develop the subjective idea that we know everything there is to know. As learners we frequently presume that we already know everything. At the same time, we tend to disregard any new information that conflicts with what has already been established.

iv. Self-Actualized Learning

The inherent motivation of students to deepen their comprehension and maximize their learning capacity is referred to as self-actualization.

v. Specialization Development

Outstanding efficiency in a certain field of knowledge that characterizes expertise, which increases the independence, autonomy and self-regulation of teach.

vi. Knowledge Creation

The beginning of mental processes related to knowledge generation is initiated by experience, reflection, and more thorough information processing.

vii. Information Organization

This meta-level of the instructional structure presents a problem in that it requires the data to be transformed into knowledge that is both meaningful and useful.

viii. Data Driven Searching

As long as they are fresh and unorganized, data is characterized as discrete and factual investigations, facts, or characters with no meaning.

ix. Concern Inspiration

The innate impulses that stimulate and fuel our drive to know and learn are thought to be enthusiasm and exploration. Interest encourages goal-setting, draws attention to events, and evaluates whether or not learners are ready to begin moving up the meta-levels of learning.

It is the initial stage of metacognitive regularities to establish and achieve sustained success for oneself and for the next generation. Teachers and scholars in all educational settings need to develop intellectual-learning skills. Furthermore, teachers might encourage students to become more independent and strategic learners by regulating their cognition.

2.5. Theory and Model of Teachers Work Efficiency

The success of the educational career, the work presentation, the school, and the learner all depend on the teachers work efficiency.

1. The Theory of Social Organization

Social organization theory is often associated with several prominent sociologists, including; Émile Durkheim, Max Weber, and George P. Murdock, but authors Valerie et al. (1991) states that cognitive process entails setting goals, evaluating the effort and skills needed to accomplish those priorities, and projecting the results. Extrinsic and intrinsic performance information sources have both been found by social psychology as significant predictors of professional efficacy and satisfaction. Extrinsic sources come from beyond the restricted work environment, but intrinsic sources are drawn directly from the ongoing task.

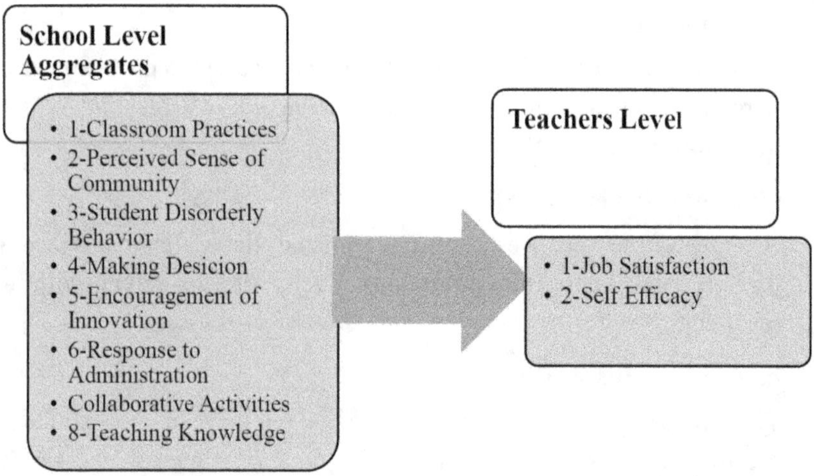

Figure No. 2.11. Social Organization Theory (Valerie, Robert, & Julia, 1991)

a. School level Aggregates

The school-level aggregates is the collective data or performance indicators that are compiled from various classrooms or departments within a school (Valerie et al., 1991).

i. Classroom Practices

Classroom practices at the school level aggregates refer to enhanced learning within individual classrooms. These practices are often evaluated collectively across the school to improve educational outcomes and ensure consistency in quality education.

ii. Perceived Sense of Community

Perceived sense of community at the school level aggregates refers to the collective feelings of connection among students, teachers, and staff within the school. It reflects the overall atmosphere of trust, inclusivity, and collaboration across the school environment.

iii. Student Disorderly Behavior

The Student disorderly behavior at the school level aggregates is the collective instances of disruptive actions, misbehavior, or violations of school rules observed across classrooms. It identifies areas that need attention for improving student behavior.

iv. Making Decision

The making decisions at the school level aggregates is the collective processes through which school leaders, staff, and stakeholders make choices and resource allocation. These decisions improve overall school performance and student outcomes.

v. Encouragement of Innovation

The encouragement of innovation at the school level aggregates refers to the collective efforts and practices within the school to foster teaching methods. It reflects the school's overall support enhance learning, teaching, and organizational practices.

vi. Response to Administration

The response to administration at the school level aggregates refers to the overall level of support between the administration and the school community.

vii. Collaborative Activities

The collaborative activities at the school level aggregates refer to the collective efforts of students, teachers, and staff working together on projects and promote a positive school culture.

viii. Teaching Knowledge

The teaching knowledge at the school level aggregates refers to the overall proficiency and effectiveness of teaching practices in supporting student learning and development.

ix. Teacher level

Teacher level mentions to the individual proficiency, expertise, and performance of a teacher in the classroom and ability to engage and support student learning.

b. Job Satisfaction

The job satisfaction at the teacher level denotes to the overall contentment and fulfillment teachers feel regarding their work and career opportunities.

c. Self-Efficacy

The self-efficacy at the teacher level denotes to a teacher's belief in their own ability to effectively manage classrooms and command on teaching challenges.

The investigation recommends that the social organization of schools has an important influence on both teachers and students through using hierarchical linear modeling techniques, the authors explore the links between school organization and the self-efficacy and job satisfaction of secondary school teachers. They also study the relationship between a teacher's sense of control over classroom practice and self-efficacy. Drawn from the Administrator and Teacher Survey from High School and beyond.

2. Organizational Commitment Theory

The commitment is the total internalized normative pressure, to act in a way that serves organizational goals and interests (Muhammad, Samina, Basharat, and Rizwan, 2010). They utilized the Organizational commitment theory, primarily attendant with John, Meyer, and Natalie (1997) they states organizational commitment as a multidimensional concept that has been understood in different ways.

Figure No. 2.12. Organizational Commitment Theory (Muhammad, Samina, Basharat, & Rizwan, 2010)

a. Organizational Commitment

The organization commitment is the employee's emotional attachment to, identification with, and involvement in the organization.

b. Job Satisfaction

The job satisfaction for teachers is a critical factor in their well-being, performance and retention in the profession with their roles, responsibilities

The job satisfaction in higher education institutions is influenced by factors and recognition of achievements. Organizational commitment, which reflects employees' loyalty and attachment to their institutions is closely linked to job satisfaction. Faculty and staff who are satisfied with their work environment are more likely to be committed to the institution's goals. Higher levels of organizational commitment lead to increased job performance. Job satisfaction and organizational commitment are essential for maintaining high-quality education and fostering a productive academic environment.

3. Teacher Cognition Theory

Gage (1964) developed the Toward a Cognitive Theory of Teaching in. This work aimed to integrate cognitive psychology into teaching practices and the school of cognitive psychology influenced Borg (1998) theory and he defines the store of beliefs, knowledge, assumptions, theories, and attitudes about all aspects of their work as the concept of language teacher cognition. Further, Tina and Mark (2012) stated that the capabilities organized by metacognitive regularities have lines seen between the three are misty and they are interdependent in practice. Teacher cognition takes an integrated approach to the processes of planning, instruction, and reflection.

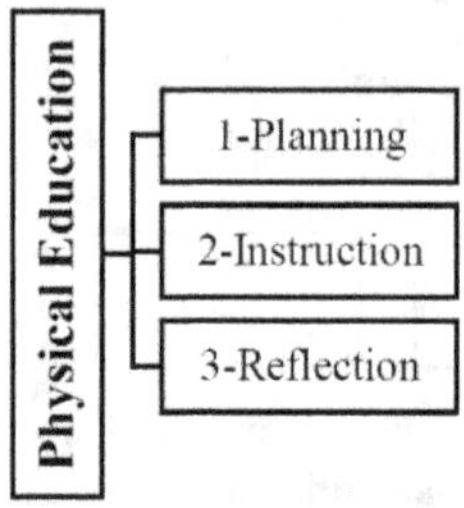

Figure No. 2.13. Teacher Cognition Theory (Tina & Mark, 2012)

a. Physical Education

The physical education refers to how teachers' beliefs, knowledge and experiences shape their teaching practices and develop students' physical literacy (Tina and Mark, 2012).

i. Planning

The planning refers to the mental processes by which teachers organize their knowledge and prior experiences. Planning in cognitive frameworks influence theirdecisions in designing effectivity and responsiveness.

ii. Instruction

The role of personal and professional experiences in shaping instructional decisions and classroom behavior.

iii. Reflection

The reflection refers to teachers' process of critically analyzing their beliefs, experiences, and teaching practices to improve classroom instruction and student outcomes.

The teaching is a multifaceted and complex process. The teacher's cognitive choices before and after instruction influence their teaching behaviors and actions. This article addresses teacher cognition in the context of physical education in teachers' preparation, instruction, and reflection, as well as in the broader area of education. Planning, instruction, and reflection have all been the subject of controlled studies up to the point.

4. Model of Work Efficiency by Siti, Sharifah, and Nik (2012)

Siti et al. (2012) state that effectiveness of the teacher is derived from projecting their professional attributes onto the performances of their teachers.

Work Efficiency

- 1-Cognititve Ability
- 2-Personality
- 3-Classroom Management
- 4-Commitment and Responsibility

Figure No. 2.14. Model of Work Efficiency (Siti et al., 2012)

a. Cognitive Ability

The way control processes regulate the various actions the brain engages in is one contentious issue in cognitive attitude. Those intellectual regulating processes setup the cognitive system to process information in a given approach, then reorganize the ability to reason when such events inform the observational participant to handle perceptions in a different way (Alejandro, 2015). Intellectual development, the ability to rationally assess two or more concepts at once and to identify when it may be beneficial to modify one's thinking and actions in response to new knowledge, is a critical attribute in self-monitoring (Donna & Marcus, 2016).

b. Personality

The university teachers' continuance performance includes not only taking part in educational events but also actively participating in research activities. Sometimes they lead to an unbalance between their individual and organizational commitments (Arifa, Mohammad, & Nazrul, 2019). The pedagogical communication, it is complex, multiphase, and multiple channels of information between two entities, individuals or groups who all at the same. They progressively conceptualized the roles of transmitters or receiver that is desirable contents in the context of the informative procedure.

c. Classroom Management

Three factors, teaching quality, interpersonal performance, and expertise can be used to describe classroom practices (Xu, Yu, Pang, Dou, & Li, 2022). Only once this initial, more fundamental distinction between correct and illusory knowledge has been made. A person's subjective expectation of accomplishing a goal or achieving a desired result through personal efforts is known as efficacy in psychology.

d. Commitment and Responsibilities

Teachers who are devoted to their educational units are more likely to do better than disinterested teachers because they are more likely to work harder on behalf of the organization and help it succeed in the process staying with the university longer (Aamir & Sohil, 2006). A good education include it as a vital component. Teachers there are committed have the love, desire, and energy they require to work harder. It is possible to confirm their commitment to the faculty by observing their propensity to meet academic goals. The roles of managers can be characterized as centers of interaction, indicating that they are responsible for facilitating communication inside an institution. It holds a few key information exchange characteristics in order to promote collaboration at the greatest level and make it more successful (Ana & Mihaela, 2018).

The appearances of cognitive ability, personality, classroom management, commitment and responsibility in participation, and continuous improvement subsequently result to good knowledge and best practices. The excellent methods of instruction enhances teacher work efficiency and devotion.

5. Organizing the Activities of a Teacher

Sergey et al. (2019) state that establishing the appropriate professional capabilities, universities and colleges will restructure the teaching process and use efficient techniques for identifying the action patterns teachers adopt during the exchange of information and abilities.

Figure No. 2.15. Organizing the Activities of a Teacher (Sergey et al., 2019)

a. Teachers Professional Activities

The teachers' professional activities in organizing their work involve planning and classroom management techniques. These activities also include ongoing professional development and reflection on teaching practices to ensure continuous improvement.

b. Motivation and Management

Motivation drives teachers' enthusiasm and commitment, while management ensures smooth implementation of teaching strategies and maintains a positive learning environment.

c. Standardization and Regulation of Teachers Work

The standardization and regulation of teachers' work in organizing activities involve the establishment of uniform guidelines and performance expectations to ensure consistency and quality across educational settings. This process align with educational standards.

The researchers investigated local regulations put in place by educational institutions as well as regulations pertaining to the activities of university teachers in order to study their professional activity and identify the extent of their work. Teachers' roles in the higher education system include stimulating students' cognitive processes and encouraging them to expand their knowledge continuously. In order to carry out this duty, instructors must also be driven, which will enable them to implement their ideas and implant them in their students. Above all, this motivation should be demonstrated by the development of professional competencies.

6. Theory of Teacher Commitment

A key factor attracting teachers' attention to learners' needs is teacher commitment. At the same time, these teachers are skilled in encouraging students to engage with the learning process (Di, 2021).

Figure No. 2.16. Teacher Commitment (Di, 2021)

a. Teacher Commitment

Regarding the direction of their work, teachers' expressive security is known as teacher commitment. It has been thought of as maybe the most fundamental element of effective teaching.

b. Promote Teaching Profession

The professional development, equitable pay, and a supportive work environment, we can acknowledge and encourage educators' devotion. This cultivates a positive, focused teaching staff that is equipped to have a long-lasting effect on students' lives.

c. Promote work Performance

It is defined as delivering high-quality instruction, positively influencing student outcomes and school success.

d. Promote School

The promoting a school involves creating a culture where teachers feel valued, supported and empowered in their roles. This commitment leads to a positive, engaging learning environment that enhances the school's reputation and student achievement.

e. Promote Student Achievement

The promoting student achievement involves fostering and dedicated teaching environment where educators are fully invested in their students' success. Committed teachers inspire, guide, and support students to reach their full academic potential.

2.6. An Overview of the Concept of Teachers' Work Efficiency's Past Evolution

Studies concerning teachers' cognitive processes since the 1970s, have emerged as a potential area of study. Richard (1973) directed an important research project on teachers' thinking, asserted that one of the fundamental abilities in any instructional act is the outcome of a decision, whether conscious or not. The Teachers' decision-making is divided into two categories based on the arrangement and participatory phases of instruction.

The teachers establish long-term judgments throughout the phase of analyzing and planning. The teacher decides in the curriculum planning as per the needs of each students as well as according to the classroom environment. Decision-making in teaching and organizational roles professionally and individually, are deliberate judgments made in the classroom based knowledge (Rasmussen, 1985). Then Killion and Todnem (1991) they expanded Schön's stages in a manner similar to these phases. Afterward, James and Peter (1993) investigated the pre-active interactive and post-active phases that teachers go through in relation to their intellectual process with its temporally related link to their classroom action. Moreover, Magdalene and Deborah (1999) added a comment mentioning to teachers as dilemma managers, who will be making decisions throughout every step of the teaching-learning process. The importance and difficulty of teachers' decisions is highly recognized since they cannot be distinguished from elements of the instructional environment.

Later on after starting of 21st century, John, Sharon, David, Karen, and Kelly (2005) argue that teachers compress their reaction to the multiple components of managing groups and education into just one move. Teachers are unable to control the multifaceted dynamic of classroom happenings. This argument is based on previous inquiry, Brown and McIntyre (1993) stated that consider problems carefully; as a result, selections in the participatory phase is fast, usual and typically unconscious. Additionally conceptualized by Koster and Dengerink (2008) the mixture of expertise, abilities, viewpoints, standards, and unique features, enabling the teacher to put into practice were. Afterward, Judit (2016) extends concept and define teacher's competence, efficiency and efficacy in certain circumstances, using them logically in European education commission.

1. Teacher Work Efficiency Social Theory v/s Social Cognitive Theory

A pair of associated educational beliefs, Rotter (1960) contributed to the growth of teacher efficacy. Moreover, Jeffrey, Pamela, and Donetta (2009) state that optimism refers to one's perception of their ability to execute a specific activity in a specific environment. In comparison to less effective teachers, he believes that effective teachers put in more efforts, utilize more problem-solving techniques, gain higher abilities, become more competent teachers, and show greater courage in the face of setbacks.

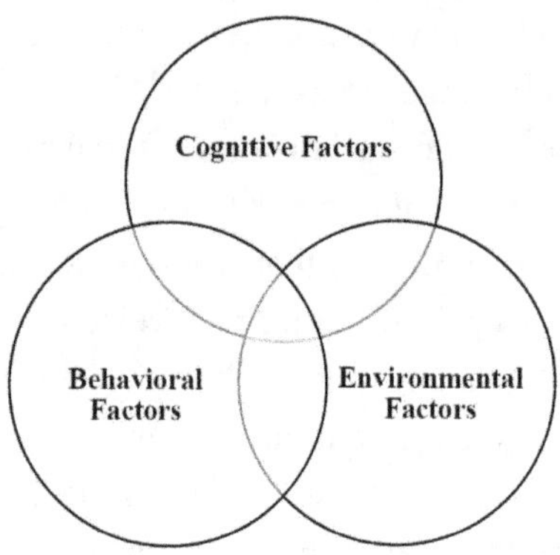

Figure No. 2.17. Social Cognitive Theory (Bandura, 1989)

Bandura (1989) model of social cognition help us understand how metacognitive regularities work with both fundamental ideas because planning, monitoring, and evaluation all require judgments and also help individuals appear more courageous when faced with challenges. Lalu and Eny (2021) define teacher effectiveness through Vigor, dedication, and absorption are qualities that define commitment to work. It stands for an upbeat and mentally satisfying state of mind. The provision of physical, emotional, informational, and instrumental support that a person experiences from his or her social circle is known as social encouragement, and social support serves a teacher's effectiveness as well.

2. The Social Cognitive Effect on Teachers Work Efficiency

There are two cognitive presumptions in the human cognitive architecture. First, it's critical to emphasize that this concept has two dimensions: educational effectiveness and individual effectiveness. The initial considerations are related to the teacher's capacity and competency to produce favorable work outcomes. The confidence in one's capacity to instruct successfully is the second dimension (Ana, Maely, Simone, Karla, & Fernando, 2016). Long- and short-term memory can be used for social context, reflection, and dealing with stored knowledge when engaging in meta-regularities in instructional environments. The capacity and duration of working memory are constrained. Small amounts of information are temporarily kept in systems with this type of memory. Another long-term memory is a type of memory that holds enormous amounts of information for a brief period of time known as cognitive loops, as opposed to working memory (Houichi & Sarnou, 2020). It is believed that people can control their individual actions through motivation, influencing situations, acts, and obstacles from the context that reflects their efficacy beliefs. The theory of social cognition established one of the few theories associated with human control that establishes a gap between ability and possibility.

3. Work Efficiency Reflects in Quality of Performance

The term efficiency is a dynamic and complex socioeconomic term that is covered by a variety of intellectual areas that serve as a foundation for conducting a multidisciplinary study of this phenomenon. The broader objective of these is the creation of practical tools for university performance management. The teacher's capacity to work in the most recent university atmosphere is an essential requirement for advancement (Sergey, Olga, & Victor, 2015). Further more numerous teaching-oriented approaches to teaching might not be sufficient. It makes sure that our learning and instruction processes increase our efficacy as teachers and are similarly beneficial when establishing curriculum and evaluations (Afzal & Md, 2021).

4. Source of Teachers Work Efficiency

Metacognition refers to a teacher's capacity to take the necessary actions to develop efficient approaches for solving the challenges they experience, determine results and implications and adjust the way they teach according to the need based on their prior expertise. Khoirotun et al. (2025) states that connection with the effectiveness is the capacity of a program, organization, or activity to accomplish its objectives. Teachers select the appropriate intellectual resource for this reason. Metacognition promotes teachers in effectively achieving a personal goal in order to increase the productivity of university teachers. Here, the sources to regulate metacognition in order to work more effectively.

a. Metacognitive Intervention

The metacognitive intervention increases the efficacy and long-term sustainability of students' acquisition of important concepts, which frequently serve as the foundational understanding for mastering more complex material. The system we use, which we call the Retrieval-Monitoring-Feedback (RMF) strategy, is based on the following two ideas: a) teacher can increase learning's resilience by practicing inquiries that are spread apart and (b) effectiveness if the timing of learning activities is well managed, instruction is based on precise performance monitoring, which includes useful feedback (Katherine & John, 2013).

b. Control and Evaluation

Katherine and John (2013) state, identifying errors, paying attention, and retrieving memories are all part of the coordination process known as metacognitive regulation. It requires interaction at the object and meta-levels of analysis. While control controls attention and employs cognitive methods, metacognitive monitoring checks teaching evaluations and feelings.

c. Knowledge-Based Beliefs

The epistemic views are conceptual frameworks that allow learners to interpret information through their own lens with more complex views permitting more adaptable comprehension. Teachers' personal conceptions of knowledge and understanding, or epistemic beliefs, have a big impact on the way they go about their metacognitive work. Teachers build more complex epistemologies as they examine the beginnings, ecosystems, boundaries, and approaches of human knowledge (Stephanie, Elmar, & Rainer, 2013).

d. Implementation of a Regulatory Process for Efficient Work

Teachers will be motivated to effectively apply if there is a thorough investigation of the issue and a more intentional formulation of their curriculum, instruction, or career goals. The adoption of MR classroom methods is seen as an example of educational inventiveness, but it was first stated as a specific goal (Jeltsen, Free, Veronique, Ankelien, Tine, & Koen, 2014).

e. Motivations behind the Regulatory Process

According to Rasha (2020) states that university teachers are focused by their pupils' academic success and their own personal performance goals. Teacher-led metacognition regulation exercises including monitoring, planning, and evaluation are used to improve memory and teaching for performance. Advanced thinking that exerts direct influence over the thought processes used in the work procedure is also referred to as metacognition. Intrinsic motivation refers to behavior that is done for institutional fulfilment rather than for an independent outcome, which has a beneficial impact on academic performance. To attain the finest pedagogical results, a teacher's performance is critically dependent on their own intrinsic motivation.

2.7. In Current Era

In this recent era, individuals engaged in work assess several aspects of their employment, it including their employer, institutional leaders, the actual work they do, and the work atmosphere. Energy levels, emotional satisfaction, effectiveness, and participation are indicators of job involvement (Jalal, 2016). The educational environments and teaching commitment were established by teachers' attitude of engagement. The years invested in teaching determine the amount of effort teachers put into advancing learners' education and well-being. Dedicated and competent university teachers are obligated to upgrade their power to manage learners' unique needs, to further build their skills (Di, 2021). Additionally, Turgut, David, Tijen, and Sedat (2025) states that two of the most important elements affecting student outcomes are leadership and professional development for teachers.

2.8. Analysis of Researches Done in the Related Fields

Numerous studies have examined how metacognitive patterns influence different aspects of academic performance. The concept of memory training, introduced by John (1971) in pioneering research on memory, has recently been explored from educational perspectives as well as in other fields of study.

Next, Flavell and Wellman (1975) stated the enhancement of cognitive processes introduced a framework for monitoring and regulation. This model emphasizes what occurs during the act of knowing and empowers individuals to take control of their own cognitive processes, enabling them to manage and improve these processes. It involves understanding metacognitive events, awareness of goals or tasks, relevant strategies, and applicable knowledge.

The categories of metacognitive knowledge include: first, declarative knowledge, which involves awareness of the subject, the task, and the strategies; and second, procedural knowledge, which pertains to the regulation of cognitive operations, evident through planning, management, and evaluation. Later, these were classified as declarative and procedural knowledge explained by Sonia and Sandra (2008) metacognitive techniques are actions aimed at understanding our activities and cognitive processes, as well as knowing how to apply them effectively in alignment with goal-oriented organizational oversight. After three decades researchers return actively to metacognition consideration. According to Farhat and Suhail (2013) metacognition refers to the ability to consciously recognize, regulate, and intentionally direct one's own thinking processes. Again, Baker, Lamoureux, Ken, and Donald (2013) state that metacognition involves being aware of the processes themselves, along with organizing, monitoring, regulating, and evaluating the steps associated with tasks and problem-solving.

Progressive organizational monitoring strengthens the understanding of goal recognition. Nosratinia and Adibifar (2014) stated that goals and quality criteria serve as benchmarks for evaluating the quality of learning. Teachers' ability to organize, guide, regulate, monitor, and assess the learning process through metacognitive regularities is a comprehensive skill that can significantly enhance students' learning effectiveness.

An investigation was conducted into the mechanisms underlying the development of the Easily Learned, Easily Remembered (ELER) rationale across three age groups of young children (4–5, 6–7, and 8–9 years). The evaluation of their progress was based on ELER reasoning, using a trial-to-acquisition approach. To systematically influence the validity of the ELER association that supports these metacognitive patterns, a new experimental design consisting of six stages a preliminary test, four training phases, and a post-test was implemented. The findings revealed that ELER judgment emerges early (around ages 4–5) but its application diminishes after implicit training (Marie, Sylvie, and Thierry, 2015).

Next year Aguirre (2016) states that metacognition is the cognitive ability that allows individuals to understand what they know, develop plans for action, remain aware of ideas during instruction, and reflect on as well as evaluate the stages and processes involved in teaching. This research concept emphasizes declarative reasoning and procedural knowledge.

The above investigation points toward the generalization of academic problems by Kinzie (2019) states that teachers' ability to apply simplifications, solve problems in context, make decisions, and work independently, collaboratively, actively, creatively, reflectively, and critically represents the core objectives of the teaching process.

Recent reviews indicate that effective teaching is approached through the consistency of teaching practices. In this context, the teacher recruitment process was selected, and three key sub-processes understanding, retention, and transformation were enhanced through the application of metacognitive regularities (Roger, Adán, & Danilza, 2020).

Additionally, Mikhail and Leona (2020) stated that another study revealed that providing explicit instructions and fostering cognitive understanding of statistical education can indirectly influence the learning of statistics; however, no research has specifically examined the metacognition of statistical concepts. The study employed a segmentation task that required dividing a continuous stream of syllables into distinct recurring units, alongside an artificial language learning paradigm. During this process, statistical learning may lead to an understanding of statistical regularities that can be reflected in the interpretation of input and discrete components. Conceptualizing items retrieved from memory as complete units, as well as novel components, is believed to represent knowledge of statistical structure.

Afterward, Daniela, Mario, Tania, Ceres, Teresa, and Pedro (2021) state that to progress from novice to proficient, software engineering students develop a combination of advanced social and cognitive skills alongside technical expertise. These challenges are better in remote learning environments due to the heightened need for autonomy and both self- and co-regulation. To address these concerns, the Sim-Programming methodology was adapted to the digital distance learning environment as e-Sim-Programming. In its second iteration, metacognitive challenges (MC) were incorporated into the e-Sim-Programming framework to foster students' reflective thinking about their learning processes.

Most recently, over time Anatoly, Elena, and Anna (2023) they conducted a research study examining the significant theoretical and practical challenges in identifying and explaining the key patterns in the development of metacognitive regulation of job-related tasks among information technology (IT) specialists during university training and the initial stages of professionalization.

Furthermore, Nikolaos, Eleni, and Maria (2023) states that over the past few decades, the education sector has increasingly focused on metacognitive regularities within quality management approaches, placing strong emphasis on international standards and various quality certifications. Additionally, Maziahtusima, Idayat, and Sakinah (2025) states that metacognitive strategies can cultivate a lifelong ability to learn. Moreover, they drive academic achievement and equip students to adapt to the rapidly changing educational and professional landscape.

An increasing number of academic publications in the field of quality studies emphasize the crucial role of teachers of all ages in universities, indicating that quality management in academia has reached a significant level of maturity. Consequently, the focus has shifted from merely procedures and resources to the core principles necessary for fostering a culture of quality and organizational excellence. Total quality management is an approach that enhances both academic and institutional performance. Significant advancements in students, teaching practices, and universities reflect an active transformation in organizational climate and pedagogical conditions.

2.9. Critical Analysis of the Literature Review of the Study on Metacognitive Regularities and Teacher Work Efficiency

The literature offered an in-depth analysis of metacognitive patterns and the effectiveness of teachers' work. Valerie et al. (1991) stated that Cognitive processes include setting goals, assessing the required effort and skills, and anticipating outcomes. Both intrinsic and extrinsic factors affect professional effectiveness and satisfaction. The study indicates that school organization impacts teacher self-efficacy and job satisfaction, with the level of control teachers have over classroom practices being a crucial factor. After a decade Muhammad et al. (2010) stated that commitment represents an internalized normative drive to align actions with organizational goals and interests. Factors such as cognitive ability, personality, managerial skills, commitment, and continuous improvement contribute to the application of effective knowledge and best practices, while strong teaching methods enhance teachers' efficiency and dedication to their work. Therefore, a comprehensive examination of metacognitive theories outlines social metacognition and effectively emphasizes its role in promoting self-regulation, problem-solving, and the collaborative benefits associated with social metacognition. However, Ming and Sze (2009) states that it presumes equal contribution from all group members, disregarding possible differences in participation and cognitive abilities.

Although the theoretical discussions clearly demonstrate metacognitive principles, their practical application across diverse educational settings still needs further empirical validation. The differentiation between metacognition, metacognitive regulation, and metacognitive knowledge remains consistent. Furthermore, Markeya (2016) focus on metacognition teaching through strategic instruction. The multiple theories and models, such as those by Rachel (2014); Dongho and Cheolil (2017), provide depth to the study. The literature highlights the importance of teacher performance in ensuring academic quality, connecting it to metacognitive patterns, regulatory processes, and overall work efficiency.

Additionally, Sergey et al. (2019) Universities and colleges will restructure their teaching processes to strengthen professional competencies. This includes implementing effective methods to assess the teaching patterns educators use when imparting knowledge and skills, as well as other related aspects. Then, Thenmozhi (2019) states that the core of this discussion is the concept of metacognition, which includes knowledge, regulation, and practical teaching applications. It provides an approach to improving teaching processes and enhancing professional capabilities.

Lisa and Richard (2020) study examined fluency, defined as temporal regularity rather than speed, as a key factor in metacognitive monitoring. It suggests that individuals rely on fluency cues to guide their metacognitive judgments, linking consistent timing to task confidence. This perspective supports the monitoring sub-variable in the current research. Furthermore, Daniel and Christian (2020) emphasize the importance of regularities in improving learning and provide steps and guidelines for incorporating metacognitive coaching into teaching practices. Their work underscores the relevance of these regularities to teaching methodologies and establishes a clear connection between work efficiency and metacognitive tools such as planning, monitoring, and evaluation.

Moreover, Di (2021) highlights the significance of teacher commitment in improving student outcomes, emphasizing the crucial role of teacher metacognition in planning, monitoring, and evaluation. (Damien et al., 2021), it does not adequately address real-world challenges such as resource limitations or the diverse needs of students. While suggestions like applying metacognitive regularities are valuable, their practical implementation across different educational contexts remains underexplored. The effect of these metacognitive regularities on enhancing teacher performance or student outcomes is also insufficiently examined. Moreover, factors such as the integration of technology into metacognitive teaching receive minimal attention, which represents a significant limitation in modern educational settings.

The literature concludes that incorporating metacognitive practices can improve educational outcomes by promoting teacher self-regulation and adaptive instructional strategies. However, further research is needed to address practical implementation challenges and explore cross-disciplinary applications to strengthen these findings. Additionally, the literature offers a foundational understanding of how metacognitive regularities influence teacher work efficiency, presenting a well-structured theoretical framework that enhances both comprehension and application.

a. Limitation of the Study

The study focused only on universities in Islamabad that were recognized by the HEC, which limits the generalizability of the findings to other regions or unrecognized institutions.

The research investigated only a one-way effect, which restricts understanding of possible reciprocal or bidirectional relationships between variables.

The study utilized only the concept of metacognitive regulation from the broader theory of metacognition, thereby limiting the exploration of other important metacognitive components.

The sample was confined to university teachers in Islamabad, which may not represent teachers from other cities, educational levels, or contexts.

The study did not examine inverse relationships between the variables, limiting the depth and comprehensiveness of the findings.

The data collection relied on questionnaires and semi-structured interviews, which may introduce self-reporting bias and limit the richness that could be obtained through additional methods such as observations or longitudinal tracking.

Chapter Three

Methods and Procedures

This chapter provides a structured perspective to fully understand the concept, as it outlines the study's methodology. The following section presents the specifics of the methodology.

3.1. Research Design

The study's design refers to the structured techniques and steps used to collect and analyze data related to the variables of the current research problem. The convergent parallel design is a mixed-methods research approach that integrates quantitative and qualitative data to provide a comprehensive understanding. In this design, researchers collect and analyze both types of data independently, then merge the results to compare findings and offer a more accurate and validated interpretation. The explanatory study addressed the objectives and null hypotheses by providing detailed explanations and interpretations. This design is particularly effective when researchers aim to validate or cross-check findings, as the two data types complement each other. Louis, Lawrence, and Keith (2018) stated that Validity is essential for research to be effective and to ensure the accuracy of the instrument's measurements. This design compares or combines results to draw a more comprehensive interpretation of the findings. It is particularly effective when researchers aim to validate or cross-check findings, as the two data types complement each other. For example, quantitative data may provide statistical results, while qualitative data offers deeper insights. Moreover, open-ended questions enhance the depth of the study by allowing respondents to provide more detailed and flexible answers. Lastly, the mixed data were integrated during the analysis. Through careful planning, this design ensures that both types of data align in terms of timing, research questions, and focus, while also allowing for equal emphasis on each during analysis and interpretation.

a. Research Paradigm:

A mixed-methods study uses both quantitative and qualitative data to investigate underlying reasons. According to Vibha and Christine (2019) Pragmatism connects reality (its effects) with epistemology (metacognitive regularity). It ties reality to epistemology by emphasizing that the truth of a belief lies in its practical consequences. This concept operates effectively in real-world situations, where outcomes matter. It is shaped by actual results in reality. In this study, pragmatism provides an operational framework for interpreting the world around us, placing strong emphasis on the practical effects of beliefs and actions. As a philosophical foundation, pragmatism prioritizes the research question and the practical consequences of inquiry over strict adherence to a single methodology or epistemological stance. To fully explore the relationship between quantifiable cognitive strategies and practical work outcomes, this study incorporates both pragmatism and positivism into the analysis of university teachers' metacognitive regularity and work efficiency. By adopting a pragmatic paradigm, mixed-methods researchers focus on what works in practice, enabling them to address problems effectively. Pragmatism endorses epistemological flexibility within the philosophy of knowledge, acknowledging that both objective (quantitative) and subjective (qualitative) perspectives can provide valuable insights. It supports integrating these approaches to offer a comprehensive understanding of the issue and informed recommendations. Therefore, Edmonds and Thomas (2017) stated that the pragmatic viewpoint directs us to analyze concepts based on their real-world applications, offering a more adaptable and flexible approach to understanding and retaining information.

b. Research Approach:

A research approach refers to the strategy used for gathering, analyzing, and interpreting study data. This mixed-methods research combines both quantitative and qualitative approaches to explore the research objectives and questions. In the triangulation approach, the researcher alternately collects quantitative and qualitative data and then compares the two datasets. This approach involves integrating both types of data, typically during the interpretation stage (Creswell, 2009). This can reduce bias and increase reliability, providing meaningful insights, while mixed methods can address both "how much" and "how many." According to Louis et al. (2018) the mixed-methods approach is characterized by a combination of diverse perspectives. This dual focus bridges the gap between specific, in-depth insights and broader applicability. Qualitative methods capture the complexity and variability of particular contexts. By integrating quantitative and qualitative data, mixed-methods research can present both detailed, participant-specific perspectives and overarching themes or trends that are common across a population. This integration allows findings from both types of data to generate insights that may not emerge when each type is examined independently. According to Edmonds and Thomas (2017) integration focuses on the conclusions or insights derived after collecting and analyzing both numerical and non-numerical data. By combining these data types, researchers can verify findings and develop a deeper understanding of the research topic. For example, quantitative data may reveal statistical trends or relationships, while qualitative data can provide contextual depth.

3.2. Data Collection

Data collection is an essential process in every research development. The data for this study was collected in 2023 from teachers at public and private universities in Islamabad. Letters of permission and research proposals were issued by the National University of Modern Languages (NUML), Islamabad, and were based on an official approval letter for all affiliated universities. The data collection process began after receiving approvals from the respective university authorities. Data was gathered through university visits and an emailed Google Form. Additionally, qualitative data was collected through semi-structured interviews. Steve (2020) stated that parametric tests rely on the assumption that the data collected is evenly distributed across the entire population. When data is collected using instruments such as Likert scales, parametric tests can be applied. Regression can be performed using models such as 50 + 8n, where n represents the total number of variables.

a. Population of the Study

The target population of this study comprised university teachers from Islamabad, Pakistan. There were 7 private and 16 public sector universities in Islamabad. In total, the 23 HEC-recognized universities from both public and private sectors had a target population of 7,743 public sector university teachers and 2,721 private sector university teachers.

Table No. 3.1.
Population of Islamabad University Teachers

Sr. No.	Strata	No. of Universities	No. of Teachers
1.	Public Universities	16	7743
2.	Private Universities	07	2721
	Total Universities	23	10464

Note. The data was received from the HEC Higher Education Statistics department of Islamabad 2021-2022.

i. Criteria for Selecting the Population

The study applied the following criteria for selecting university teachers: public and private sector HEC-recognized universities in Islamabad. University teachers were chosen as participants for both qualitative and quantitative components. The study did not categorize participants based on gender, department, or online teaching restrictions; therefore, every university teacher was considered a respondent. Teachers from all faculties were included according to the sample size. The two major strata private and public sector universities were divided into two groups. The number of private-sector teachers was smaller than that of public-sector teachers. To determine an appropriate quantitative sample for each stratum, 5% of teachers from each stratum were selected as the quantitative sample for the study.

b. Sampling Techniques

A sampling technique is used to select a subset of individuals from a larger population to make inferences about the entire population (Gay, mill, & Airasian, 2012). The study objectives focused on subgroups of university teachers, such as those from public sector and private sector universities in Islamabad.

i. Quantitative Sampling Technique

In this study, variables were quantified, and findings were generalized across the population using the proportionate stratified sampling technique. Proportionate stratified sampling is a statistical method in which the population is divided into distinct subgroups, or strata, based on shared characteristics. A sample is then drawn so that the size of each stratum in the sample is proportional to its size in the overall population. This approach ensures that each subgroup is effectively represented in the sample in the same proportion as it exists in the population. Consequently, this technique enhances the representativeness of the sample, leading to more accurate and reliable inferences about the population (Gay et al., 2012). In the current study, the population consisted of 10,464 university teachers from HEC-recognized universities in Islamabad.

Table No. 3.2.
Distribution of Sampling

Public and Private sector University Teachers	Total	Public Sector	Private Sector
Population	10464	7743	2721
Sample size (5%) in number	524	388	136
Division of sample size (5%) in percentage	5%	3.7%	1.3%

Table No. 3.2. explain the distribution of sample size which was 5% of the 10464 or 524 university teachers. So, taking 5 % from each statra, the 1.3% sample taken from the private sector, which was 5% of the private sector universities population and the 3.7% sample was from the public sector, which was the 5% of the public sector universities population. The ratio is different for private and public university teachers because public university teachers are more in numbers.

ii. Qualitative Sampling Technique

Qualitative data were collected in parallel with quantitative data, and the qualitative responses were supported by the quantitative survey. The researcher employed a purposive sampling technique to obtain the qualitative sample. Bruce (2001) stated that after conducting fieldwork on a specific group, samples were selected. Then, Edmonds and Thomas (2017) stated that a qualitative data-gathering phase-specific sample generation was ensured for the respondents. Afterwards, Louis et al. (2018) states that the purposive sampling technique provides depth and was chosen for a specific purpose: to access knowledgeable individuals for an improved study. It was used to obtain a qualitative sample. Creswell (2018) states that the purposive sampling was further linked to homogeneous sampling, which allows the selection of respondents with identical characteristics and traits. Twenty-one semi-structured interviews were conducted to explain the criteria. The researcher sampled 15 university teachers from public universities and 6 university teachers from private-sector universities.

c. Sample Size

i. Quantitative Sample size

The researcher selected 5% of the population to represent the entire population, thereby reducing the use of resources and time. The population was divided into private and public university teachers using proportionate stratified sampling. The table below also shows the sample return rate from university teachers.

Table No. 3.3.
Sample of the Quantitative Study

Sr. No.	Universities Sector	No. of Universities	Total No. University Teachers	Total *n* (05%)	No. of Return	Rate of return in Percentage
1.	Public Sector Universities	16	7743	388	338	87%
2.	Private Sector Universities	7	2721	136	115	100%
3.	Total no of Universities	23	10,464	524	453	86%

ii. Qualitative Sample size

Purposive sampling was used to select individuals who could provide in-depth information on the subject through a qualitative approach (Complete Dissertation, n.d.). A structured interview was conducted with 4% of the teachers from both sectors. This 4% qualitative sample was drawn from the 5% quantitative sample. A total of 15 university teachers from the public sector and 6 teachers from private sector universities were selected from 23 universities in Islamabad.

Table No. 3.4.
Sample of the Qualitative Study

Sr. No.	Sector Universities	No. of Universities	Total No. of University Teachers	Total *n* (05%) of quantitative	Total n (04%) of 524
1.	Public Sector Universities	16	7743	388	15
2.	Private Sector Universities	7	2721	136	6
3.	Total no of Universities	23	10,464	524	21

d. Tool Construction

The investigation utilized instruments to gather quantitative data. The hypotheses were tested using quantitative results obtained through the closed-ended questionnaire (Burke and Larry, 2014). The closed-ended questionnaire was developed as a two-scale instrument. A semi-structured interview sheet was used for conducting the interviews. The instrument began with demographic questions to collect information about the university teachers. The conceptual framework selected for the study represented teachers' work efficiency and metacognitive regularities. The first scale of the questionnaire was developed through metacognitive regularities model given by (Julie et al., 2021). The second scale of the questionnaire was created on the work efficiency model presented by Siti et al. (2012) The Teachers' Work Efficiency Scale consists of four subcategories, while the Metacognitive Regularities Scale comprises three subcategories. Overall, the instrument was divided into three sections: demographics, Scale 1 (MRS), and Scale 2 (WES). A semi-structured interview sheet was also used to explore the practices of university faculty members. An explanation of the research tools, along with an outline of the demographics, is provided below:

i. Demographics

The first section of the tool collected information about the respondents' demographics. It focused on parameters such as type of university, experience, job status, and designation. This information provided a basic overview of the respondents' backgrounds.

ii. Metacognitive Regularities Scale

Initially, 90 closed-ended items were created for both scales of the questionnaire. The first scale, which measured metacognitive regularity, included 43 closed-ended questions covering three subdomains: planning, monitoring, and evaluation.

Table No. 3.5.
List of Items (Initial Version) Teachers' Metacognitive Regularities Scales (MRS)

Scale	Major Selections	No of Items	Item Coding
Teachers' Metacognitive Regularities	Planning	15	PL1 – PL15
	Monitoring	14	M1 – M14
	Evaluation	14	E1 – E14

According to Table No. 3.5. there were 15 items in the Planning segment, ranging from PL1 to PL15 code No., in the Monitoring segment 14 items ranging from M1 to M14 code No. as well as 14 items in the Evaluation segment with E1 to E14 code No.

iii. Work Efficiency Scale

Similar to the first scale, 47 closed-ended items were previously created for the second scale. Its four subdivisions were cognitive abilities, personality, classroom management, and commitment and responsibility.

Table No. 3.6.
List of Items (Initial Version) Teachers' Work Efficiency Scales (WES)

Scale	Major Selections	No of Items	Item Coding
Teachers' Work Efficiency	Cognitive Ability	09	CA1 – CA9
	Personality	16	P1 – P16
	Classroom Management	09	CM1 – CM9
	Commitment and Responsibility	13	CR1 – CR13

Table No. 3.6. shows that there were 09 items in the cognitive ability segment ranging from CA1 to CA9 code No., in the personality segment 16 items ranging from P1 to P16 code No., in the classroom management segment 9 items ranging from CM1 to CM9 code No., as well as 13 items in the Evaluation segment with CR1 to CR13 code numbers.

iv. Five Point Likert Scale

The Likert scale is a system of measurement used to quantify attitudes, behaviors, and opinions (Katherine & Kimberly, 2017). The Likert scale is well-suited for capturing respondents' levels of agreement. This study utilized five-point scales in the quantitative questionnaire;

Strongly Disagree (SD)

Disagree (D)

Neutral (N)

Agree (A)

Strongly Agree (SA)

v. Semi-Structured Interview Sheet

A semi-structured interview sheet was used for an in-depth study. This qualitative study was conducted through semi-structured interviews with teachers, and the interview sheet included 27 open-ended questions.

vi. Validity of the Instrument

The extent to which a questionnaire measures what it is intended to measure in an investigation is known as validity (Roberta & Alison, 2015). The drafted Metacognitive Regularities Scale and the Work Efficiency Scale in this study assessed metacognitive regularities and work efficiency within the same questionnaire. The tool was validated by four experts before data collection. These experts provided suggestions for improvement based on their knowledge, with the goal of making the tool more effective in gathering reliable data. The conceptual framework was used to design the scales of the questionnaire. Both scales were reviewed by specialists in light of the study's objectives, title, and conceptual framework. A total of 90 questions were drafted for the two scales, which were based on closed-ended questions. After incorporating the experts' comments and recommendations, the questionnaires were revised, restructured, and finalized. Improvements were achieved by following the insightful feedback of experts in the field of education. The list of corrections is given below:

Table No. 3.7.
Instrument Validity Details of Metacognitive Regularities Scale (MRS)

Item Code.	Item No.	Old Items	Revised Items
PL2	2.	I plan my pedagogical task the week before	I plan my pedagogical task a few days before for class.
PL4	3.	I organize my career developmental path by planning.	I plan activities for my career development.
PL5	4.	The result of my lesson preparation are reflected in class interaction.	I feel lesson planning helps me in a better way.
PL10	5.	I feel I can compensate for my weak point with proper plans.	I can compensate for my weak point with proper plans.
PL13	6.	I plan subject objectives to improve my course efficiency.	I plan course objectives to improve my course efficiency.
M3	7.	I observe, the best way to fix topics in student minds is to do a discussion.	I observe the best way to teach topics is to do a discussion.
M4	8.	I have noticed that monitored classes give the best grades in universities.	I noticed that progress-tracked classes give the best grades in universities.
M10	9.	I believe that monitoring is essential for curriculum development.	I believe impact monitoring helps to improve the curriculum.
E12	10.	I know policy challenges for teachers in university.	I always face policy challenges brilliantly at university.

Table No. 3.8.
Instrument Validity Details of Work Efficiency Scale (WES)

Item Code	Item No.	Old Items	Revised Items
CA1	11.	I use my cognitive ability in the planning of classroom teaching.	I use verbal reasoning in classroom teaching.
CA4	12.	I usually utilize my cognitive abilities to avail difficult career opportunities.	I utilize my working memory to avail career opportunities.
CA5	13.	I examine any subject thru cognitive capacity.	I examine any subject thru visual processing.
CA7	14.	I use critical abilities to understand my students as a teacher.	I use social cognition to understand my students as a teacher.
CM1	15.	I maintain silence in classroom management.	I manage silence in classroom for discipline.

Table No. 3.9.
Instrument Validity Details of Section 1. Metacognitive Regularities Semi-Structure Interview Sheet

Q No.	Old Items	Revised Items
12.	Does evaluation help to organize the best work?	Does evaluation help to enhance the quality of work? How?
16.	Are you agree that monitoring effects on teaching?	Do you agree that monitoring effects teaching? Why?

Table No. 3.10.
Instrument Validity Details of Section 2.
Work Efficiency Semi-Structure Interview Sheet

Q No.	Old Items	Revised Items
9.	Does an effective teacher organized atmosphere in the university?	Is the university environment well-organized by the teachers? Why?
3.	Is the teacher's attitude makes effects on the teaching and learning process?	Does the teacher's attitude affect the teaching and learning process? How?

Table No. 3.7, 3.8, 3.9 and 3.10. All two instrument validity tables have 4 columns. Column 1 has item codes, and the 2nd column has the serial no. of the question. Items before editing correction have been listed in the third column. Items after correction have been listed in the fourth column.

vii. Pilot Testing of the Study

A pilot study was conducted to determine the reliability of the quantitative instrument. According to Julie (2001) the reliability of the scales depends on the sample used. Pilot testing confirmed their trustworthiness with the specific study sample. According to Roberta and Alison (2015) measuring reliability refers to the ability of a tool to consistently produce similar results from individuals completing a motivation assessment on multiple occasions. The tool was initially distributed to 52 respondents from the sample, of which 38 submitted complete responses, resulting in a return rate of 72%. The data collected during the pilot trial were coded and analyzed using the 20th version of the Statistical Package for the Social Sciences (SPSS).

viii. Reliability of the Instrument

To determine the effectiveness of the scale and refine the items for the final version of the tool, the data collected during the test phase were analyzed. For this purpose, item-total correlation, inter-section correlation, and Cronbach's alpha were applied to compute reliability.

Table No. 3.11.
Cronbach Alpha Reliability of Teachers' Metacognitive Regularities Pilot Testing Scale (n=38)

Scale	Major Section	Items	Cronbach Reliability Alpha
Metacognitive Regularities		43	0.959
	Planning	15	0.906
	Monitoring	14	0.873
	Evaluation	14	0.879

Table No. 3.11. Displayed that the reliability of the teacher metacognitive regularities scale (MRS) was 0.959. While the reliability of the planning section 12 items showed 0.906 indicating excellent reliability, monitoring section 10 showed 0.873 demonstrate strong internal consistency with a Cronbach's alpha of 0.873 and evaluation 9 items showed 0.879 reflecting a high level of reliability. Overall, reliability for the total scale and its subs-variables are well.

This indicating that the scale is a reliable measure of teachers' metacognitive regularities.

After collecting the complete data, the researcher again checked the Cronbach Alpha Reliability of Teachers' Metacognitive Regularities Scale Final Tool (n=453).

Table No. 3.12.
Item-total Correlation of Teachers' Metacognitive Regularities Scale Pilot Testing (n=38)

Items Codes	r	Items Codes	r	Items Codes	r
PL1	0.777**	M1	0.555**	E1	0.612**
PL2	0.698**	M2	0.621**	E2	0.600**
PL3	0.663**	M3	0.605**	E3	0.737**
PL4	0.686**	M4	0.600**	E4	0.600**
PL5	0.633**	M5	0.719**	E5	0.591**
PL6	0.711**	M6	0.574**	E6	0.674**
PL7	0.636**	M7	0.545**	E7	0.737**
PL8	0.524**	M8	0.605**	E8	0.586**
PL9	0.404**	M9	0.478**	E9	0.578**
PL10	0.638**	M10	0.697**		
PL11	0.604**				
PL12	0.685**				

Correlation is significant at the level 0.01 level (2-tailed). **

Correlation is significant at the0.05 level (2-tailed).*

Table No. 3.12. Specifies the item-total correlation of the Metacognitive Regularities Scale (MRS). The item-total correlations for the planning items ranged from r = 0.404 to r = 0.777 showed strong correlations. The item-total correlations for the monitoring items ranged from r = 0.478 to r = 0.719, showed a moderate correlation and the evaluation items ranged from r = 0.578 to r = 0.737 exhibited strong correlations. The highest correlation was of item No. PL1 (0.777**) and the lowest correlation was of item No. PL9 (0.404**). Overall, indicate that the majority of items within each domain are strongly related to the total score, supporting the reliability and internal consistency of the scale in this pilot sample.

After collecting the complete data, the researcher again checked the Item-Total Correlation of Teachers' Metacognitive Regularities Scale (n=453).

Table No. 3.13.
Inter-section Correlation of Techers' Metacognitive Regularities Scale Pilot Testing (n=38)

	Planning	Monitoring	Evaluation	TMR
Planning	1			
Monitoring	0.878**	1		
Evaluation	0.861**	0.906	1	
T M R	0.961**	0.962**	0.954**	1

Correlation is significant at the level 0.01 level (2-tailed). **

Correlation is significant at the 0.05 level (2-tailed).*

Table No. 3.13. displayed that all the sections were correlated with one another significantly based on a pilot test with 38 participants. The results indicate a strong positive correlation between Planning and Monitoring (0.878), as well as between Planning and Evaluation (0.861). Additionally, Monitoring is strongly correlated with Evaluation (0.906), these high correlations suggest that the components of the TMR scale are closely related and contribute significantly. Overall, the maximum correlation was found between monitoring and teachers' metacognitive regularities (0.962) while the lowest correlation was found between planning and evaluation (0.861).

After collecting the complete data, the researcher again checked the Inter-section Correlation of Techers' Metacognitive Regularities Scale (n=453).

Table No. 3.14.
Cronbach Alpha Reliability of Teachers' Work Efficiency Scale Pilot Testing (n=38)

Scale	Major Section	Items	Cronbach Reliability Alpha
Work Efficiency		47	0.983
	Cognitive Ability	09	0.924
	Personality	16	0.947
	Classroom Management	09	0.921
	Commitment and Responsibility	13	0.964

Table No. 3.14. indicates that the reliability of the teacher work efficiency scale, consisting of 42 items and alpha=0.983, shows excellent internal consistency for the whole scale. The reliability of the sub-variables, cognitive-ability, which consists of 09 items alpha=0.924 suggests high reliability. Personality consists of 11 items and alpha=0.947, demonstrating high reliability, classroom management consists of 09 and alpha=0.921, indicating strong internal consists and lastly, commitment and responsibility consists of 13 items and alpha=0.964, reflecting excellent reliability.

After collecting the complete data, the researcher again checked the Cronbach Alpha Reliability of Teachers' Work Efficiency Scale (n=453).

Table No. 3.15.
Item-total Correlation of Teachers' Work Efficiency Scale Pilot Testing (n=38)

Items Codes	r	Items Codes	r	Items Codes	r
CA1	0.725**	P6	0.855**	CM9	0.502**
CA2	0.571**	P7	0.877**	CR1	0.885**
CA3	0.841**	P8	0.799**	CR2	0.809**
CA4	0.696**	P9	0.775**	CR3	0.806**
CA5	0.818**	P10	0.855**	CR4	0.778**
CA6	0.731**	P11	0.698**	CR5	0.807**
CA7	0.740**	CM1	0.758**	CR6	0.875**
CA8	0.778**	CM2	0.654**	CR7	0.892**
CA9	0.681**	CM3	0.703**	CR8	0.730**
P1	0.621**	CM4	0.804**	CR9	0.622**
P2	0.699**	CM5	0.756**	CR10	0.762**
P3	0.648**	CM6	0.824**	CR11	0.802**
P4	0.799**	CM7	0.840**	CR12	0.830**
P5	0.776**	CM8	0.681**	CR13	0.839**

Correlation is significant at the 0.01 level (2-tailed). **

Correlation is significant at the 0.05 level (2-tailed).*

Table No. 3.15. identifies the item-total correlation of the Work Efficiency Scale range from the highest correlation of r=0.892 to the lowest r=0.502, indicating varying degrees of relationship strength between each item contribute meaningfully to the scale's overall score, with stronger correlations reflecting more consistent alignment with the total scale. The cognitive ability items have ranged from r=0.571 to r = 0.841. The personality items have ranges from r = 0.621 to r = 0.877. The classroom management items have ranged from r = 0.502 to r = 0.840. The commitment and responsibilities items have ranged from r = 0.622 to r = 0.892. These results indicate that the majority of the items show strong correlations with the total score, which supports the reliability of the scale in measuring teachers' work efficiency.

After collecting the complete data, the researcher again checked the Item-Total Correlation of Teachers' Work Efficiency Scale (n=453).

Table No. 3.16.
Inter-section Correlation of Techers' Work Efficiency Scale Pilot Testing (n=38)

	Cognitive Ability	Personality	Classroom Management	Commitment and Responsibility	Teachers Work Efficiency
Cognitive Ability	1				
Personality	0.808**	1			
Classroom Management	0.846**	0.830**	1		
Commitment and Responsibility	0.807**	0.897**	0.944**	1	
Teachers Work Efficiency	0.896**	0.943**	0.957**	0.976**	1

Correlation is significant at the 0.01 level (2-tailed). **

Correlation is significant at the 0.05 level (2-tailed).*

Table No. 3.16. suggested that strong, positive correlations among all the measured variables ranged from 0.807 to 0.976. The relationships between cognitive ability, personality, classroom management, commitment and responsibility, and teachers' work efficiency were strong. Cognitive ability was (r =0 .896), indicating that higher cognitive ability. Personality shows a strong positive correlation with (r =0 .943). Classroom management is also positively correlated with (r = 0.957), demonstrating that better classroom management skills are associated with higher efficiency in teachers' work. Commitment and responsibility exhibit the highest correlation with (r =0 .976), implying that a strong sense of commitment and responsibility is crucial for maximize work efficiency. Overall, Commitment and Responsibility and teachers' Work Efficiency were shown to be highly correlated (0.976**) but the correlation between cognitive ability, and commitment & responsibility was the weakest (0.807**). These results show a correlation in enhancing teachers' work efficiency.

After collecting the complete data, the researcher again checked the Inter-section Correlation of Techers' Work Efficiency Scale (n=453).

e. Research Study Ethics

The British Educational Research Association [BERA] (2018) and the American Psychological Association (2017) guidelines on research ethics were followed in this investigation. In this study, it was ensured that participants were fully informed about the study's objectives, methods, and benefits before providing consent to participate.

University teachers had the freedom to decide whether or not to participate. Participants' privacy was protected by ensuring that their responses and personal information remained confidential. During data collection, participants' identities were kept anonymous by avoiding the inclusion of any identifying information. University teachers also retained the right to withdraw from the study or refuse to provide responses at any time, without concern, as a way of respecting their autonomy and rights.

It was ensured that respect for teachers was maintained and that the moral standards of the university were upheld. The research procedures were explained transparently to avoid any conflicts of interest in the findings. The responses from university teachers were safeguarded against unauthorized access, use, or disclosure. Email communication and data collection procedures were conducted in accordance with institutional ethical standards. Finally, appreciation was expressed to every teacher for their voluntary participation in the research.

3.3. Data Analysis

Descriptive analysis involves summarizing and interpreting data to gain insights. It focuses on providing a clear understanding of the variables through measures such as mean, median, and mode. The analysis in this study focuses on the linear regression relationship with the dependent variable. Linear regression was applied to the dependent variable based on practical information related to the independent variable. A formal statistical test was conducted to assess the normality assumption. The null hypothesis was evaluated using the specified level of significance (Schuyler, 2012). The study uses normally distributed statements along with the null hypothesis. Descriptive analysis enables researchers to identify potential areas of interest, detect irregularities, and establish a foundation for further analysis. Inferential analysis, on the other hand, goes a step beyond descriptive analysis by using sample data to make inferences or generalizations about a larger population. This includes techniques such as regression analysis. Descriptive and inferential analyses are often interconnected, with descriptive analysis providing the basis for inferential analysis. Before conducting an inferential analysis, a descriptive analysis is typically performed to examine the distribution of the data.

Although mix methods incorporate both the philosophical principles of deductive and inductive analysis. Jared, Brain, and Lea (2014) state that Descriptive statistics provide valuable frequency information and insights about the data. Tests such as regressions are essential for informed decision-making. Descriptive analysis explains the “what” of your data, while inferential analysis addresses the “why” and “how,” enabling broader conclusions and predictions from sample data. Both types of analysis are essential, but inferential analysis typically builds upon the insights gained from descriptive analysis.

a. Quantitative Data Analysis

The explanatory analysis leads study to the final stage of our journey, where simple bivariate analysis involves the use of independent and dependent variables. Therefore, Norman (2003) states that the most basic form of explanatory analysis uses a symmetrical coefficient to indicate the degree of statistical influence one variable has on another. One or two research questions may require some form of associational analysis. Bivariate descriptive analysis serves two main purposes. Moreover, Creswell (2015) states that Descriptive statistics helped summarize the overall trends in the data. Quantitative data were tested using SPSS version 20 to evaluate the responses collected through the questionnaire. The regression between the dependent and independent variables in a two-variable model is referred to as linear regression. A more advanced step involves multiple regression, where a dependent variable may be influenced by multiple independent variables. Steve (2020) states that an effective size of $r = 0.1$ is the lowest, $r = 0.3$ is the average, and $r = 0.5$ is the greatest. R values of 1%, 9%, and 25% account for the whole variance. The statistical test that was used for testing the hypotheses is described in the table 3.17.

b. Qualitative Data Analysis

After coding and reviewing the qualitative data, similar themes emerged. Bruce (2001) states that the analysis of qualitative data combines explicit coding procedures with a theory development approach through an analytic process of constant comparison. Continuous relevance, along with joint coding and analysis, explains the theory more systematically and with greater analytical clarity. Thematic analysis was applied in this study. Creswell (2015) states that Themes are clusters of similar codes aggregated to represent a major idea in the database. They form a core element in qualitative data analysis, and thematic analysis connects these themes to theoretical and conceptual models. The interpretation of qualitative findings was then related to the quantitative findings. The qualitative coding used for data analysis is presented in Table 3.18.

Table No. 3.17.
Data Analysis of the Study

Sr. No.	Objective	Research Questions	Null Hypotheses	Data analysis techniques	Assumption
1.	To investigate the metacognitive regularities in university teachers	What are the metacognitive regularities of University teachers?		Mean	
2.	To determine the work efficiency of university teachers.	What is the level of the work efficiency of university teachers?		Mean	
3.	To assess the Effect of metacognitive regularities on work efficiency of university teachers.	How do university teachers consider the effect of their metacognitive regularities on their work efficiency?	There was statistically no significant effect of metacognitive regularities on work efficiency of university teachers.	Linear Regression	Normality of distribution validates statistical tests of the regression coefficient
3a.	To assess the effect of planning on work efficiency of		There was statistically no significant effect of	Linear Regression	Normality of distribution validates statistical

	university teachers	planning on work efficiency of university teachers.		tests of the regression coefficient
3b.	To assess the effect of monitoring on work efficiency of university teachers.	There was statistically no significant effect of monitoring on work efficiency of university teachers.	Linear Regressi on	Normality of distributio n validates statistical tests of the regression coefficient
3c.	To assess the effect of evaluation on work efficiency of university teachers.	There was statistically no significant effect of evaluation on work efficiency of university teachers.	Linear Regressi on	Normality of distributio n validates statistical tests of the regression coefficient

Table No. 3.18.
Depiction of Themes and Sub-Themes with Coding Scheme

Sr. No.	Main Theme	Sub Theme Categories	Coding Scheme
1.	Metacognitive Regulatory	*1. Planning*	i. *Planning comprise*
			ii. *Planning effect on students' performance*
			iii. *Planning improve course efficiency*
			iv. *Planning complete commitment*
			v. *Planning schedule work in advance*
		2. Monitoring	vi. *Monitoring organize best work*
			vii. *Monitoring difficult to do*
			viii. *Presentation help student work monitoring*
			ix. *Achieving step by step commitment*
			x. *Class resources updated*
		3. Evaluation	xi. *Provide factual support*
			xii. *Evaluation enhance the quality of work*
			xiii. *Evaluation promotes subject inspection*
			xiv. *Evaluation found gap in teaching process*
			xv. *Evaluation consider annual practice*

2.	Work Efficiency	*1. Cognitive Ability*	i. *Cognitive ability helps in teaching practice* ii. *Cognitive ability helps in class room management*
		2. Personality	iii. *Personality affect the teaching* iv. *Personality is the combination of attitude*
		3. Classroom Management	v. *Silence is effective for classroom management* vi. *Group activity is useful in classroom management*
		4. Commitment and Responsibility	vii. *Performance intervened the commitment* viii. *Commitment increase your responsibility*
3.	Effect of University Teachers Metacognitive Regularities on their Work Efficiency	*1. Metacognitive Regulatory Effect on Work Efficiency* *2. Work efficiency Determination*	xvi. *Enhance teaching* xvii. *Integrates the information* ix. *University environment remain well organized by teachers work.* x. *Teachers affects students over all well-being*

Chapter Four

Analysis and Interperetation of the Data

4.1. Summary of Analysis

The fourth chapter provides an in-depth review of the respondent profile, research objectives, research questions, hypotheses, and the statistical tests used, including mean and regression analysis. It also presents a descriptive analysis of the data collected through a semi-structured interview sheet and questionnaire, along with an interpretation of the results. The following five sections provide further subdivisions of the data analysis.

a. Demographics

The first section of this chapter provides a general introduction to the respondents, including various percentages based on their designation, job status, and experience. Although demographic information was not the primary focus of this study, it was necessary for identifying the faculty members.

b. Descriptive Statistics

Descriptive statistics offer a summary of the data collected from the study's sample, which can then be generalized to the entire population. Relevant tests are included as part of the procedure. Therefore, in an investigation, quantitative descriptions are presented through the use of descriptive statistics (William, James, & Kanika, 2016).

c. Objective 1 (Metacognitive Regularity of University Teachers)

The chapter presents the initial goal. The mean test score is used to address Objective 1, which is to investigate the metacognitive regularities among university teachers. The first research question provides qualitative data to support the qualitative arguments. The hybrid approach, known as a concurrent combination design, involves collecting data separately and interpreting each data point together.

d. Objective 2 (Work Efficiency of University Teachers at the University)

When developing new and improved environmental projects, mixed methods are especially beneficial. The mean test score is used to address Objective 2, which is to determine the work efficiency of university teachers. Research Question 2 asks: How can the work efficiency of university teachers at the university be determined? The second question in this chapter also supports a mixed-methods research study.

e. Inferential Statistics

In this section, the conclusions drawn for the population were analyzed. Hypothesis testing was used in this investigation (Creswell, 2015). Moreover, William et al. (2016) states that the integration of inferential statistics helps to draw conclusions about the broader population from the sample data.

f. Objective 3 and Sub-Objectives 3a, 3b, 3c (Effect of Metacognitive Regularities on Work Efficiency of University Teachers)

The third goal and its sub-objectives are addressed in this section. A regression test was used to conduct the analysis. Objectives 3, 3a, 3b, and 3c were designed to evaluate the effects of metacognitive regularities and their sub-variables, such as planning, monitoring, and evaluation.

g. Thematic Analysis

The technique used for analyzing qualitative data was thematic analysis, which involves collecting data and identifying themes through the search for patterns of meaning.

h. Research Question 1. And 2. (Q1- What are the metacognitive regularities of university teachers? Q2- How to determine the work efficiency of university teachers at the university? and Q3- How do university teachers consider the effect of their metacognitive regularities on their work efficiency?)

To obtain insider viewpoints, semi-structured interviews were conducted with university teachers. Each teacher was asked 27 questions derived from the conceptual framework. The respondents (teachers) were assigned alphabetical codes by the researcher to protect their privacy. A total of 21 university teachers from HEC-recognized universities in Islamabad participated in the qualitative study. The semi-structured interviews were conducted through in-person visits to the teachers.

i. Comparison of Results

This section presents the main construct for the convergent parallel design, which combines quantitative and qualitative data. The study design involves integrating both types of data after their independent analysis to produce a comprehensive investigation of the study's variables, leading to important conclusions (Creswell, 2009).

4.2. Demographics

Table No. 4.1.
The Sector-wise Respondent University Teachers' Distribution (n=453)

Universities type	Frequency	Percentage
Public	359	79.2%
Private	94	20.8%

Table No. 4.2.
The University Teachers' Distribution as per the Job Status (n=453)

University Teachers Designations	Frequency	Percentage
Lecturer	256	56.5%
Assistant Lecturer	103	22.7%
Associate Professor	84	18.5%
Professor	10	2.2%

Table No. 4.3.
The University Teachers' Distribution as per the Designation (n=453)

University Teachers Job Status	Frequency	Percentage
Permanent	271	59.8%
Contracting	90	19.9%
Visiting	92	20.3%

Table No. 4.4.
The University Teachers' Distribution as per the Experiences (n=453)

University Teachers Experiences	Frequency	Percentage
0-2 years	77	17.0%
2-5 years	99	21.9%
5-10 years	143	31%
10-20 years	122	26%
20+ years	12	2.6%

4.3. Descriptive Statistics

a. Metacognitive Regularity of University Teachers

Objective 1: To investigate the metacognitive regularities in university teachers

Table No. 4.5.
The Metacognitive Regularity of University Teachers (n=453)

Sr. No.	Variable	Sub-Variable	*n*	Mean of Mean	Status
1.	Metacognitive Regularities		453	3.97	Agree
		Planning	453	4.07	Agree
		Monitoring	453	3.96	Agree
		Evaluation	453	3.84	Agree

Table No. 4.5. Indicates the metacognitive regularities in university teachers of Islamabad. The mean score of metacognitive regularities was 3.97. The planning mean score was 4.07. The monitoring mean score was 3.96 and the evaluation mean score was 3.84. The result showed that planning regarding the work efficiency of respondents was agreed upon planning for the best learning environment, a few days before planning pedagogical tasks, planning activities for career development, lesson planning, marking planned commitments on the calendar, conducting student activities by managing class time, recognize teaching strength with help of planning, plan helps in lecture formation, planning compensate weak points, helps in planning course objectives and planning helps in targeted result in university working are agreed.

The result displayed that monitoring concerning the metacognitive regularities of respondents was agreed upon, Teachers track lesson plans to improve lecture delivery, teachers observe that discussion gives the best result for teaching, class progress depends on monitoring class records, presentations are the best way to check student progress, and monitoring makes it possible for teachers to review their professional development, tracking helps teachers to accomplish course responsibilities, monitoring helps to fulfill teaching commitments, monitoring gives best effect on curriculum, class resources are updated by monitoring and teachers agreed that universities assigned task are always followed up for the best future result.

The result shows that evaluation regarding the metacognitive regularities of respondents was agreed upon that teachers consider course evaluation annually, classroom observation is the key concern for teaching, evaluation improves teaching, evaluation found gaps in teaching, monitoring detect deficiencies in student learning and teaching task evaluate. Teachers agreed that they evaluate themselves through teaching task committedly, topic evaluation promotes subject evaluation, teachers face policy challenges brilliantly through evaluation and teachers knew that institutional standard established on evaluation.

b. Work Efficiency of University Teachers

Objective 2: To determine the work efficiency of university teachers.

Table No. 4.6.
Work Efficiency of University Teachers at the University (n=453)

Sr. No.	Variable	Sub-Variable	*n*	Mean of Mean	Status
1.	Work Efficiency		453	3.91	Agree
		Cognitive Ability	453	3.96	Agree
		Personality	453	3.99	Agree
		Classroom Management	453	3.77	Agree
		Commitment and Responsibility	453	3.91	Agree

Table No. 4.6. determined the Islamabad university teachers' effectiveness. University teachers' work efficiencies of mean score was 3.91. The cognitive ability of mean score was 3.96. Personality of mean score was 3.99. Classroom management mean score was 3.77. Commitment and Responsibility mean score was 3.91.

The results of this study demonstrated that respondents' cognitive ability and work efficiency were in acceptance, verbal reasoning is useful for pedagogies, intellectual capacities promote classroom instruction, teachers utilize working memory to avail career opportunities, teachers examine the subject through visual processing, utilize conceptional skills for managing task, utilize social cognition to understand the students and they utilize intellectual capacities to take decisions for students.

The results of the investigation showed that respondents' work efficiency and personalities were viewed as acceptable that teachers believe to be a popular teacher among students, believe strong relationships with students and teaching go hand in hand, they believe that using strict discipline will not change students' behavior, they apply all the qualities in teaching to help the students, they agree that behavior creates characters, they control their mood swings in the classroom, they are controlling their feeling in difficult situations, they investigate the scenario before taking class decisions, they consider that monitoring makes it easy to advise students, they tackle work missions carefully and they keep versatility in their personality for their colleagues.

The results demonstrated that respondents' work efficiency and classroom management were appropriate, they agree on managing silence in the classroom, generally, they utilize a teamwork strategy in class, they are in favor of students sitting in straight lines, they respect university regulation in the classroom, they like to design classroom with a focus on the students, they help pupils regularly to motivate them, they utilize only the administration's provided resources in the classroom and they constantly ask the student to keep the classroom tidy.

The results of this study indicated that the work efficiency, commitment, and responsibilities of the respondents were effective, teachers completed the subject curriculum with responsibility, they wisely met the demands of the students' learning, they engaged every learner in period, teachers took part in all activities of the institution, they take responsibility for building up students' social skills, they believe that teachers maintain academic standards, they trust that the interpersonal abilities of students are developed by teachers, teachers are always a part of the university establishment workgroup, they spend time on updating university policies, they always offer a hand to other officials as a good university colleague, they evaluate classrooms with an accurate approach, they respect their seniors' comments heartily, teachers are committed to carrying out the classes at the appointed hour.

4.4. Inferential Statistics

a. Effect of Metacognitive Regularities on Work Efficiency of University Teachers

Objective 3: To assess the effect of metacognitive regularities on the work efficiency of university teachers.

$H_0 1$: There is statistically no significant effect of metacognitive regularities on work efficiency of university teachers.

Table No. 4.7.
Effect of Metacognitive Regularities on Work Efficiency of University Teachers (n=453)

Independent Variable	Dependent Variable	R^2	β	t	Sig.
Metacognitive Regularities	Work Efficiency	0.831	0.415	5.512	0.00

a. Independent variable = Metacognitive Regularity
b. Dependent variable = Work Efficiency
**p<0.01, *<0.05

Table No. 4.7. presents the results of a regression analysis that considers the effect of metacognitive regularities on the work efficiency of university teachers. The value of R^2 =0.831 indicates that 83.1% variance in work efficiency can be explained by metacognitive regularities. This suggests a strong explanatory power of the independent variable on the dependent variable. The value β=0.415 signifies a positive relationship between metacognitive regularities and work efficiency. The t-value of 5.512 is a measure of the statistical significance of the relationship between the independent and dependent variables. A high t-value indicates a significant relationship. The significance value ($p < 0.01$) is 0.00, which is below the 0.01 threshold. This indicates that the effect of metacognitive regularities on work efficiency is statistically significant at the 1% level. The null hypothesis was fail to accept. This means that there is a statistically significant effect of metacognitive regularities on the work efficiency of university teachers. Therefore, metacognitive regularities positively and significantly influence the work efficiency of university teachers.

i. Effect of Planning on Work Efficiency of University Teachers

Objective 3a: To assess the effects of planning on work efficiency of university teachers.

$H_0 1a$: There is statistically no significant effect of planning on work efficiency of university teachers.

Table No. 4.8.
Effect of Planning on Work Efficiency of University Teachers (n=453)

Sub-Independent Variable	Dependent Variable	R^2	β	t	Sig.
Planning	Work Efficiency	0.756	0.869	9.183	0.000

a. Independent variable = Planning
b. Dependent variable = Work Efficiency
**p<0.01, *<0.05

Table No. 4.8. displayed that The R^2 value of 0.756 indicates that 75.6% of the variance in work efficiency of university teachers can explained by their planning. This is a strong indication that planning affects work efficiency. The Beta coefficient of 0.869 suggests a strong positive relationship between planning and work efficiency. This implies that as planning increases, work efficiency also significantly increases.

The t-value of 9.183 is significantly high. The significance level (p-value) of 0.000 is less than 0.05, which means the result is statistically significant. Therefore, the null hypothesis failed to be accepted. This indicates that the effect of planning on the work efficiency of university teachers is statistically significant. It is concluded that planning, as measured by metacognitive regularities, has a positive effect on the work efficiency of university teachers.

ii. Effect of Monitoring on Work Efficiency of University Teachers

Objective 3b: To assess the effects of monitoring on work efficiency of university teachers.

$H_0 1b$: There is statistically no significant effect of monitoring on work efficiency of university teachers.

Table No. 4.9.
Effect of Monitoring on Work Efficiency of University Teachers (n=453)

Sub-Independent Variable	Dependent Variable	R^2	β	t	Sig.
Monitoring	Work Efficiency	0.760	0.872	7.967	0.000

a. Independent variable = Monitoring
b. Dependent variable = Work Efficiency
**p<0.01, *<0.05

Table No. 4.9. presents the results of a regression analysis investigated the effect of monitoring on the work efficiency of university teachers. The value of R' 0.760 indicates that approximately 76% of the variance and monitoring is a strong predictor of work efficiency. The standardized β = 0.872 suggests a strong positive relationship between monitoring and work efficiency. The t-value of 7.967, coupled with a significance level ($p < 0.001$), indicates that the effect of monitoring on work efficiency is statistically significant. The null hypothesis failed to be accepted. This finding suggests that monitoring, as measured by metacognitive regularities, has a statistically significant effect on the work efficiency of university teachers.

iii. Effect of Evaluation on Work Efficiency of University Teachers

Objective 3c: To assess the effects of evaluation on work efficiency of university teachers.

$H_0 1c$: There was statistically no significant effect of evaluation on work efficiency of university teachers.

Table No. 4.10.
Effect of Evaluation on Work Efficiency of University Teachers (n=453)

Sub-Independent Variable	Dependent Variable	R^2	β	t	Sig.
Evaluation	Work Efficiency	0.706	0.840	9.694	0.000

a. Independent variable = Evaluation
b. Dependent variable = Work Efficiency
**p<0.01, *<0.05

Table No. 4.10. demonstrate that 70.6% of the variance, evaluation significantly predicts work efficiency, as indicated by the high-value $R^2=0.706$ and a strong standardized $\beta = 0.840$. The $t= 9.694$ is significant at the 0.001 level. These results suggest that evaluation has a statistically significant positive effect on the work efficiency of university teachers. Therefore, the null hypothesis failed to be accepted.

4.5. Thematic Analyses of Qualitative Data

a. Metacognitive Regularities

Objective 1. To investigate the metacognitive regularities in university teachers.

Research Question 1. What are the metacognitive regularities of university teachers?

Themes Emerged

Twenty-one university teachers were interviewed. The purpose of the semi-structured interviews was to investigate the effect of metacognitive regularities on university teachers through a variety of questions. Semi-structured questions were asked during the interviews. The systematic procedure of categorizing qualitative feedback and combining the codes that subsequently serve as variables of interest for the researcher in comprehending a phenomenon is known as thematic analysis. To protect the teachers' privacy, the researcher replaced their identities with alphabetical codes. In Section 1 of the research instrument, the researcher developed five questions based on each indicator, and in Section 2, two questions were formed for each indicator. The themes, sub-themes, and sub-sub-themes are presented in Table No. 3.18.

i. Planning

The first theme identified through thematic analysis was planning. The interviewees mentioned that planning involves having a clear vision aligned with the mission, positively impacts student performance, improves course efficiency, ensures commitment through the use of a calendar, and emphasizes the importance of scheduling daily work activities in advance. The university teachers stated that

1) Planning Comprise

A, highlighted the importance of understanding the cause, which support initial concept in planning. B, expressed confusion regarding planning supports. However, C, said that having a clear vision is crucial for effective planning. This claimed was supported by D, who emphasized that planning itself is rooted in vision and mission. E, repeated he same idea by stating that vision and mission serve as the targets of planning efforts. F, contributed by describing planning as a pathway toward upcoming achievements. G, Supported the concept that planning helps in accomplishing tasks aligned with missions and visions. H, determined the discussion by underlining that planning cannot be successful without a clear vision and mission. I, emphasized the significance of knowing the mission, which supports the planning's visualized concept. J, spoke concern about the planning support. K, stated that good planning requires a clear vision. L, underlined that purpose and vision are the foundation of planning itself. M, claiming that planning efforts are directed on the vision and mission. N, helped by establishing planning as a means of achieving future goals. O, accepted that planning facilitates the completion of tasks that are connected. P, promoted the value of the initial planned concept. Q, stated concern about the planning encouragement. R, claimed that efficient planning needs a clear vision. S, stated that the foundation of planning is vision and mission. T, saying that planning efforts are directed toward the vision and mission. Finally, U, explaining the process of planning as a means of achieving foreseeable objectives.

2) Planning Effects on Students Performance

A, agreed for the significance of planning. B, added by focusing its relevance in the context of lesson planning, indicating its role in structured education delivery. Further on this concept C, suggesting that planning creates rules and helps students in staying on track without conscious effort. D, responds by admitting the unevenness in teachers' grip on timelines and the necessity of planning to change this. E, emphasized the connection between planning and student engagement, underscoring its role in fostering active participation in learning activities. F, emphasized the confidence it instills in teachers, suggesting that well-prepared teachers are more effective in their teaching roles. G, discussed the idea of planning as providing guiding channels, showing its role in directing the learning process. H, highlighted the share of development, suggesting that organized planning by teachers helps in students' goal-setting abilities. I, acknowledged the need of planning. J, Significance in planning highlights its function in offering of planned instruction. K, making the argument that planning helps learners remain on directions. L, acknowledges the inequity of teachers' control over deadlines and the need for preparation to address this. M, indicating that instructors who are well-prepared are more successful in their teaching positions. N, demonstrated that teachers who are well-prepared are more successful in their positions. O, Planning serves as a guiding framework that directs and shapes the learning process. P, highlighted the importance of planning. Q, Planning's importance focuses on its role in providing planned teaching. R, stated that planning helps pupils stay on course and establishes rules. S, replies by acknowledging that teacher control over schedules is varied and that planning is required to fix this. T, highlighted that they are more successful in their teaching positions when they are well-prepared. Lastly, U, indicated that instructors who are organized conduct their jobs better.

3) Planning Improve Course Efficiency

A and B, agreed, highlighting about planning enhancement process that leads to better outcomes. However, C and D, offered different views, noting that planning may not always be effective since it's fundamental to the course itself and efficiency in the end depends on results. E and F, contributed by pointing out the benefits of planning, such as selecting suitable words and examples or ensuring quality material for each topic. Lastly, G and H, emphasized the practical aspects of planning, suggesting preparing topics before class and ensuring continuity by linking information through topics. I, improving our planning methods is crucial because it directly contributes to achieving stronger and more successful results. J, focus on enhancing how we plan, it sets us up for smoother execution and more positive results. K, argued that planning isn't always effective, as it's just a foundation, and real efficiency is judged by outcomes. L, suggesting that planning alone doesn't guarantee efficiency outcomes matter more. M, Highlighted the value of planning by choosing the right words, examples, and quality content for each topic. N, emphasized planning's role in selecting effective words, examples, and strong content per topic. O, Focused on effective planning through pre-class topic preparation and connections. P, refining the planning process plays a key role in driving better outcomes and long-term effectiveness. Q, enhancing the planning phase helps streamline processes and consistently leads to improved performance outcomes. R, pointed out that while planning is essential, its success is ultimately measured by the results it delivers. S, noted that planning may not always lead to efficiency, since results are the true indicator of success. T, Showed how planning helps in using appropriate language, examples, and ensuring high-quality material. Lastly, U, demonstrated that planning leads to better words choice relevant examples, and well prepared materials.

4) Planning Complete Commitment

A and B, showed the value of calendars in recalling tasks and serving as a planner itself. C, expanded on this, saying about calendars marking for reminders, helping in organization and task prioritization. D, emphasized the importance of dedication in making planning effective, suggesting that commitment is vital for successful execution. E, adds on this by showing the need of setting clear deadlines within a plan. F, offered a practical approach, suggesting the creation of a priority list and its integration into the planner. G, identified the significance of managing activities within identified timeframes, supporting with effective planning strategies. H, determined the discussion by pointing out the usefulness of calendar alarms in reminding individuals of their commitments before of time. I, demonstrated how calendars help remember tasks and act as planners. J, showed calendars as tools for both reminders and planning. K, setting calendars reminders helps structure work and ensures key tasks aren't over looked. L, noted that success depends on committed execution of well-made plans. M, enhancing our planning process with clear deadlines ensures we stay on track and achieve better results. N, suggested making a priority list and including it in the planner. O, highlighted how proper planning ensures tasks are completed within set timelines. P, highlighted calendars' role in task recall and planning. Q, proved calendars useful for tracking and organizing tasks. R, with calendar alerts, it's easier to stay organized and manage tasks by priority. S, highlighted that commitment is key to turning plans into success. T, effective planning, especially with well-defined deadlines, leads to stronger, more predictable outcomes. Lastly, U, recommended integrating a priority list into the planning process.

5) Planning Schedule Work in Advance

A and B, both supported the tasks to be completed before planning, A, emphasized its priority in the planning process, while B, added that it should be done according to necessity. However, C, disagreed, pointing out that due to a hectic schedule, it's not always feasible to complete tasks earlier. D, supported A and B's viewpoint, equating work planners with task completion. E and F, joined in, stating that completing tasks earlier confirms a smoother workflow and minimizes distractions. G, agreed, highlighting the effectiveness of this practice in time management. H, conveyed the idea of completing tasks before planning enhances work efficiency. I, suggested the need to complete tasks ahead of planning, highlighting their importance. J, highlighted completing tasks first as essential to effective planning. K, differed, explaining that with such a packed schedule, finishing tasks ahead of time isn't always possible. L, saying that using planners directly contributes to getting tasks done. M, getting things done ahead of time helps maintain a steady flow and reduces breaks. N, early task completion ensures better focus and a more organized workflow. O, agreed, the effectiveness of this practice in time management. P, stated task completion as a key step before planning begins. Q, emphasized that finishing tasks early is crucial for successful planning. R, citing a demanding workload as a reason why early task completion isn't always practical. S, emphasizing that effective planning tools led to successful task execution. T, finishing tasks early keeps the process running smoothly and limits unexpected distractions. Lastly, U, it supports a seamless process and cuts down on disruptions.

ii. Monitoring

Respondents also mentioned that monitoring helps organize the best work, although it is difficult to do. They stated that presentations assist in monitoring students' work, monitoring supports achieving step-by-step commitment to tasks, and all class resources are kept updated through monitoring.

1) Monitoring Organize Best Work

A, emphasized the importance of focusing only on task completion rather than constant monitoring. B, acknowledged that occasional monitoring might be necessary. In comparison, C, strongly favored the concept of continuous monitoring, stating its absolute necessity. D, sought a more in-doubt approach, suggesting doubts about the benefits of monitoring. E, highlighted the value of regular feedback provided through monitoring activities. F, emphasized that monitoring helps in identifying areas for improvement and enhancement. G, focused on the role of monitoring in error detection and correction. H, identified the importance of monitoring in reviewing the effect and effectiveness of one's work. I, highlighted that our priority should be on achieving results. J, it may be important to carry out periodic monitoring from time to time. K, ongoing monitoring is absolutely vital. Without it, you risk missing critical issues as they arise. L, targeted in favor strategy, asking questions regarding the advantages of monitoring. M, emphasized the importance of regular input from monitoring activities. N, underlined that monitoring supports in determining points in need of growth and enhancement. O, focusing on the function of monitoring in identifying and fixing errors. P, pointed out that efficiency improves when the focus remains on outcomes, not on excessive oversight. Q, agree that some level of monitoring could be required occasionally to ensure everything stays on track. R, it's not optional; it's a necessary part of effective management. S, placed some doubt on the advantages of monitoring by taking a more critical viewpoint. T, showed how important it is to receive constant input from efforts to monitor. Lastly, U, highlighted the fact that monitoring contributes in finding components that require progress and enhancement.

2) Monitoring Difficult to do

A, pointed out common ground that monitoring is really important. B, however, introduced a critical viewpoint and suggested that the necessity of monitoring depends on what exactly is being monitored. C, supported the importance of monitoring by highlighting its role in maintaining a comfortable and calm environment. D, acknowledged the effort required for effective monitoring, indicating its significant feature. E, contributed by emphasizing the significance of planning, suggested that poor planning can lead to ineffective monitoring sessions. F, drew attention to the possible effects of insufficient monitoring, such as receiving incorrect data due to misconceptions in the lecture. G, added another thing by suggesting that the difficulty in monitoring may stem from an ineffective approach rather than the activity itself. H, showed the high level of skill and attention required for monitoring, suggested that it cannot be performed unconsciously. I, highlighted the importance of monitoring as a basis for collaboration. J, arguing that the need for monitoring is contingent upon the specific subject of the observation. K, emphasizing how it helps to keep a peaceful and comfortable atmosphere. L, highlighted the work necessary for efficient monitoring, highlighting its important aspect. M, contributed by highlighting the importance of planning and implying that inadequate planning may result from. N, brought to light the likely effects of weak monitoring, such as achieving insufficient information as a result of inattentive lecture. O, suggesting that an inefficient strategy, rather than the activity itself, might be the cause of the monitoring challenges. P, identified an agreed-upon view that monitoring is crucial. Q, proposed that the need for monitoring would vary depending on the specific subject being watched. R, monitoring is to preserving a relaxed and peaceful atmosphere, which reinforced its significance. S, recognized that effective monitoring requires effort, highlighting its important aspect. T, added by highlighting the importance of planning and pointing out that insufficient preparation might result in inefficient monitoring sessions. Finally, U, addressed some of the effects of inadequate monitoring, including getting incorrect information as a result of classroom mistakes.

3) Presentation Help Student Work Monitoring

A, suggested that presentations have limited benefits, while B, claimed that presentations are always beneficial. C, supported the idea about bullet points in presentations contribute to students' deeper understanding. D, agreed, highlighting the importance of students conveying their own comprehension. E, stated that presentations help improve students concern. F, added to the discussion by stating that presentations enhance communication skills. G, supported this view, emphasizing that presentations boost confidence levels. H, differences, saying that presentations might be premature for students at this stage in their academic journey. I, stated that the benefits of presentations are minimal. J, stated that presentations are always helpful. K, accepted the concept that using bullet points in presentations helps students grasp concepts more deeply. L, accepted it is for pupils to express student understanding. M, claimed that students' concerns are improved by presentations. N, Furthered, states that communication skills are improved by presentations. O, confirmed this concept by highlighting that presentations enhance confidence levels. P, represented that the advantages of presentations are low. Q, argued that learners always benefit from presentations. R, acknowledged that incorporating bullet points in presentations enhances students' understanding of the lecture. S, emphasized the need of students expressing their own understanding, sharing. T, believed that presentations aid in raising scholars awareness. However, U, contributed that presentations improve the ability to communicate.

4) Achieving step by step commitment

A, they agreed that monitoring is truly a gradual process, B, emphasized its role in facilitating informed decision-making for future actions. C, they also acknowledged that monitoring serves as a means to ensure progress through step-by-step procedures. D, contributing to the refinement of goals and keeping focus on what truly matters. E, additionally, they highlighted the accountability aspect of monitoring. F, recognizing its role in keeping individuals responsible for their work. G and H, furthermore, they emphasized how monitoring provides the necessary direction and awareness essential for successful goal attainment. I, acknowledged that monitoring is actually a slow process. J, highlighted monitoring helps people make well-informed decisions on what to do in the future. K, additionally, they recognized that monitoring is a way to guarantee progress through methodical approaches. L, helping to focus attention on the important things and improve goals. M, emphasized the responsibility component of monitoring. N, acknowledging its function in sustaining students for their studies. O, highlighted, monitoring gives the required guidance. P, stated that monitoring is actually a slow procedure. Q, highlighted monitoring assists in making better choices for next steps. R, likewise they agreed that monitoring is a way to assure success through systematic measures. S, assist in improving objectives and paying attention on what's important. T, the responsibility element of monitoring was also pointed out. However, U, being aware of its purpose in keeping individuals accountable for their job.

5) Class Resource Updated

A, suggested the importance of highlighting any deficiencies in resources, which was recalled. B, emphasized the alignment of provided resources with requirements. C, contributed for using the role of monitoring in ensuring resource sufficiency, a sense that was supported. D, suggested to update class resources. E, highlighted the user's role in identifying resource needs. F, emphasized the role of teachers in informing about the physical condition management of material resources, a point challenged by. G, by saying that such responsibility lies with management. However, H, expressed agreement with the importance of teachers' observation skills in resource management. I, emphasized the need to point out any shortages in resources, which was remembered. J, highlighted the available resources matched the specified requirements. K, the act of monitoring played a role in maintaining sufficient resources, a point that was acknowledged. L, recommended revising the materials used in class. M, emphasized the importance of the user's input in determining what resources are required. N, stated that the importance of teachers in communicating how to manage the physical state of material resources was highlighted, though this point faced opposition. O, stated that responsibility for this rests with the management team. P, replied the significance of identifying any resource gaps, which was noted. Q, focused the compatibility between the resources offered and the work needs. R, using monitoring to ensure adequate resources was a contribution that received support. S, advised making updates to the educational resources for the class. T, underlined the user's responsibility in recognizing the necessary resources. Finally, U, replied that teachers play a key role in providing guidance on maintaining the condition of physical resources, a claim that was contested.

iii. Evaluation

During the interview, respondents indicated that evaluation provides factual support, helps enhance the quality of work, promotes the subject's development, identifies gaps in the teaching process, and that course evaluation as an annual practice in academic institutions is necessary.

1) Provide Factual Support

A, believes that evaluation is indeed necessary, while B, keep the opposing view, arguing that it's not required if executed properly. C, offered a complex viewpoint, suggested that evaluation was important to somewhat, specifying a range of 60 to 70%. D, disagreed with A and firmly stated that evaluation was not necessary at all. E, expressed agreement with A, emphasizing that true evaluation provided genuine support. F, added to the pro-evaluation stand and claimed that it prevented mistakes. G, highlighted the importance of evaluation by stating that it clarified the complexities of the workflow and H, contributed by emphasizing that evaluation offered a visualization of the current state of affairs. I, thinks that assessment is truly essential. J, argue that proper execution reduces the need for it altogether. K, indicated that evaluation held moderate importance, falling within the sixty% range. L, opposed A's view and asserted strongly that there was no need for any evaluation. M, highlighting that the honest assessment offered real encouragement. N, incorporated into the pro-evaluation stand, with claims that it helped avoid errors. O, emphasized that evaluation was crucial in shedding light on the intricate nature of the workflow. P, holds the view that conducting evaluations is important. Q, states that if implemented effectively, the requirement may be resolved. R, replied the importance of evaluation was considered to be somewhat significant, estimated between sixty seven%. S, disagreed with A and insisted that conducting an evaluation was completely unnecessary. T, viewpoint that the sincere feedback gave meaningful reassurance. Lastly, U, added it to the evaluation setup, declaring that it reduced the chance of mistakes.

2) Evaluation Enhance the Quality of Work

A, expressed a viewpoint and emphasized the role of efforts in evaluation. B, agreed. C, contributed that evaluation identifies both weak points requiring improvement and strong points that offer encouragement. D, re-stated that evaluation covers both good and bad aspects of work. E, added another perspective by stating that accuracy in evaluation enhances overall productivity. F, expanded on this idea by suggesting that evaluation goes beyond simple work efficiency. G, emphasized the role of evaluation in reducing work-related problems. H, suggested that evaluation also served to prevent critical situations. I, shared an opinion and highlighted the importance of effort in the assessment process. J, it was proposed that evaluation played a role in avoiding potential crises. K, evaluation helps highlight areas that need improvement while also recognizing strengths that provide motivation. L, the evaluation includes both the strengths and weaknesses of the work. M, highlighting another viewpoint, it was noted that precise evaluation contributes to improved overall efficiency. N, further developed the concept by proposing that assessment involves more than just measuring productivity. O, the importance of assessment in addressing job-related challenges. P, conveyed a perspective while stressing how crucial effort is in making evaluations. Q, evaluation was also seen as a means of averting serious issues. R, through evaluation, we can identify shortcomings that need attention as well as strengths that serve as a source of encouragement. S, the assessment addresses both the positive and negative elements of the work. T, from a different angle, it was suggested that greater accuracy in assessments boosts general productivity, and U, elaborated on the idea by emphasizing that evaluation encompasses more than mere operational efficiency.

3) Evaluation Promote Subject Inspection

A, states that inspection is the initial step, followed by subject development. B, that's true. C and D, contribute by emphasizing that strong results confirm the subject development, aligning with the group's agreement on the importance of these steps for maximized student learning. E, stated the necessity of topic inspection and subject development for academic achievements. F, highlights that inspection enhances subjects' knowledge with more information, while G, again states that inspection promotes overall improvements. H, concludes the discussion by emphasizing that inspection, raises awareness of critical aspects. I, assessment comes first, and is then followed by the development of the subject. J, highlighting inspection, the discussion brings important issues to light and increases awareness of key factors. K, strong outcomes confirm the progress made and demonstrate the group's collective confidence in these measures to improve student learning. L, encouraging performance supports the subject's growth and underscores the shared conviction within the group about the effectiveness of these approaches to student learning. M, emphasized the importance of thoroughly examining the topic and developing the subject matter to attain academic success. N, inspection develops individuals' understanding by providing them with additional insights. O, inspections leads to broader improvements in performance and quality. P, stated that process begins with inspection, which is subsequently followed by subject development. Q, focusing on inspection within the discussion helps shed light on vital concerns and enhances understanding of crucial elements. R, positive results affirm the subject's advancement and highlight the team's united trust in these strategies to boost student learning. S, reinforce the development of the subject and reflect the group's consensus on the value of these actions in enhancing learning outcomes. T, highlighted the need for careful evaluation of the topic and structured subject development as key factors for academic accomplishment, and U, offering more detailed information, inspection contributes to a broader knowledge of the subject.

4) Evaluation Found Gap in Teaching Process

A, claims that evaluation effectively identifies missing steps. B, emphasizes that it exposes gaps that arise when problems are overlooked. C, added to the discussion by highlighting that evaluation contributes to progress by evaluating the efficacy of the plan in place. D, points out the role of evaluation in setting targets, supporting the view that it serves as a compass for advancement. E, contributes by stating that evaluation serves as a mean to recall and address missing concepts, similar to fixed forgotten points. F, strengthened this idea by likening evaluation as an act of making corrections of overlooked aspects. G, expands the scope of the discussion by stating that evaluation defines the direction of the process. H, emphasized the critical role of evaluation in analyzing and addressing instances of weak performance, underlining its importance for further improvement. I, argues that evaluation highlights omissions that occur when certain problems are ignored. J, highlights the shortcomings that emerge when issues are ignored. K, contributed to the discussion by emphasizing that evaluation plays a key role in advancing progress by assessing how effective the current plan is. L, evaluation plays a crucial role in guiding goal-setting, acting as a directional tool for progress. M, emphasizes that evaluation helps in identifying and recovering overlooked ideas, acting like a tool to revisit forgotten concepts. N, comparing evaluation to the process of identifying and correcting previously unnoticed elements. O, evaluation plays a pivotal role in shaping the course of the entire process. P, evaluation reveals the gaps caused by failing to address specific issues. Q, replied that flaws that develop from neglecting problems. R, replied discussion underlined that progress is supported through evaluation, as it helps determine how well the existing plan is working. S, informing the process of setting objectives, evaluation functions as a guiding force that steers development forward. T, suggests that through evaluation, one can revisit and fill in gaps in understanding, much like retrieving previously forgotten information. Lastly, U, the concept was supported by describing evaluation as a means of addressing and amending aspects that had initially been overlooked.

5) Evaluation Consider Annual Practice

A, focused particularly on academic progress. B, emphasized the mandatory nature of these evaluations within the course/ curriculum. C, emphasized their role in setting objectives. D, suggested reviewing them. E and F, highlight that evaluations offer numerous suggestions. G, supported performance management. H, considered it essential for making informed decisions to progress move forward. I, placed special emphasis on educational advancement. J, stated that requirement for these evaluations as a compulsory part of the course or curriculum. K, highlighted their contribution to defining goals. L, stated that crucial for making informed choices to ensure continued progress. M, assessments provide a wide range of recommendations. N, analyses generate multiple constructive ideas for improvement. O, assisted in managing employee performance. P, concentrated primarily on improving academic performance. Q, highlighted that these evaluations are an essential and mandatory component of the curriculum. R, argued their involvement in establishing targets. S, considered vital to make well-informed decisions in order to advance further. T, evaluations yield several valuable suggestions. Lastly, U, feedback from evaluations includes numerous helpful proposals.

These results highlighted the importance of vision and mission in planning, their role in education, and their impact on completing courses on time. They also emphasized the role of calendars in effective planning and time management, as well as the general view that completing tasks before planning can be a beneficial and practical approach though opinions differed on its feasibility in all situations. Additionally, the discussion underscored the importance of monitoring in work processes, the effectiveness of presentations in student learning, the benefits of monitoring in achieving goals, and the allocation and oversight of resources. Participants shared diverse viewpoints on the complexities and significance of monitoring in various aspects of their work, including resource allocation and accountability. Furthermore, the importance of evaluation in their work was acknowledged, while its challenging role in productivity was also emphasized. They agreed on the pivotal role of topic inspection and subject development in the learning process. The discussion also addressed the significance of annual evaluations within academic courses, recognizing their function as a comprehensive summary of practices. Finally, the discussion highlighted the complexity of monitoring in education, focusing on its impact, evaluation criteria, and the intricate nature of evaluation methodologies.

b. Teachers Work Efficiency

Objective 2: To determine the work efficiency of university teachers.

Research Question 2. What is the level of the work efficiency of university teachers?

Themes Emerged

i. Cognitive Ability

The first theme identified after thematic analysis is the cognitive ability of university teachers. Respondents considered cognitive ability as an important factor that supports teaching and believe it plays a significant role in classroom management. The university teachers also mentioned this aspect in their responses

1) Cognitive Ability Helps in Teaching Practices

A, yes, the teacher needs to give answers. B, said, yes, of course. C, says yes, strongly required in teachers. D, answered yes they easily understand teaching. E, states yes, helps in executing topic content. F, said yes, easily prioritize the task. G, stated yes, easily explore ideas for explaining topics. H, also said, yes, help in giving topic explanations. I, the teacher is required to provide responses. J, stated that expected to help by giving explanations of the topics. As the teacher, I am required to respond. K, confirms that it is absolutely essential for teachers. L, find it easy to grasp what's being taught. M, Confirms positively and contributes to delivering the topic content. N, yes, the task can be prioritized with ease. O, yes, it is easy to explore ideas for issues that need explanation. P, it is necessary for the teacher to offer answers. Q, replied providing topic explanations is part of my role as a teacher, it's my responsibility to respond. R, endorses it's highly necessary in educators. S, understand the lessons without difficulty. T, agrees and assists in carrying out the topic material and U, admitted, arranging the job is easy.

2) Cognitive Ability Helps in Classroom Management

A, stats that it is definitely mandatory, same as B and C, supported that expresses willingness to explore new methods. D, emphasized the importance of employing different approaches for explaining complex

concepts, while E, highlighted their effectiveness in addressing classroom issues. F, added to the discussion by affirming that diversified teaching strategies enabled teachers to offer tailored guidance to students. G, pointed out that such methods contribute to maintaining student focus during lessons, same as H, emphasized their role in fostering student interest and engagement. I, considering that it is clearly required. J, encouraged with an honesty to trying innovative approaches. K, assisted in a way that shows interest in experimenting with different strategies. L, highlighted the significance of using varied methods to clarify intricate ideas. M, emphasized that handled challenges in the classroom. N, emphasizing that using varied teaching methods allowed educators to provide personalized support to their students. O, highlighted as effective in sustaining student concentration during class. P, indications that it is surely essential. Q, supported with a readiness to accept new techniques. R, supported alternative solutions. S, replied the value of adopting multiple strategies to make complex concepts easier to understand. T, analyzed ability to resolve problems encountered in the classroom. Lastly U, highlighting how a range of instructional strategies helped teachers deliver customized assistance to learners.

ii. Personality

The second important theme was teachers' personality, which affects the teaching and learning process. Personality is considered a combination of attitudes. The university teachers also highlighted this aspect in their responses

1) Personality Affects the Teaching

A, believes that personality does surely reflect, while B, agrees but adds that it's not the only defining factor. C, shares a strong belief in the effect of personality on teaching effectiveness. D, underscores this point by suggesting that teachers' experiences shape their personalities and therefore, their teaching methods. E, highlights the role of observation, noting that students learn from the behavior and approach of their teachers. F, strengthens this concept by stating that a teacher's personality becomes evident during classroom interactions. G, states that organized personalities are crucial for achieving academic success. H, states differences from this viewpoint and suggests that personality might not be the only determinant of effective teaching. I, convinced that one's personality is inevitably revealed. J, agrees, but emphasizes that it's not the sole factor that defines the situation. K, firmly believes that a teacher's personality greatly influences how effectively they teach. L, teacher's personality and ultimately their approach to teaching is influenced by their personal experiences. M, emphasizes the importance of observation, pointing out that students absorb lessons through watching their teachers' conduct and methods. N, highlighted teacher's personality naturally reveals itself through their interactions in the classroom. O, organized is essential for succeeding in academic tasks. P, holds the view that personality undeniably shows through. Q, noting that other important elements also play a role. R, states that teaching success is significantly shaped by the educator's personality. S, view that educators' methods in the classroom are shaped by the life experiences that mold their character. T, observation plays a key role, as students often model their learning on how their teachers act and engage with their work. However U, states supported by the observation that classroom interactions reflect a teacher's true personality.

2) Personality is the Combination of Attitude

A, acknowledged the influence of situations on responses, indicating that responses are context-dependent. B, supported this view simply with a "yes." C, added complexity by suggesting that even in similar situations, responses can vary based on individual beliefs. D, supported this concept by highlighting that diverted attitudes can emerge in different situations. E, expands the discussion to include teachers and states that their thoughts and actions reveal a blend of personalities. Further F, links attitude to behavior and personality management, showing that attitude shapes teachers interactions. G, supports the views that individual personality expresses during performance moments. Finally, H, drowns a distinction between personality and attitude and states that they are separate entities. I, recognized that responses vary based on the situation, showing they depend on context. J, accepted that the nature of replies is influenced by the surrounding situation. K, stated that reactions in similar scenarios can still vary due to differences in individual perspectives and values. L, emphasizing that people's attitudes may shift depending on the circumstances. M, replied mindset and behavior reflect a mix of different personalities. N, teachers' behavior and personality influence their interactions, highlighting how attitude plays a key role. O, suggestion that a person's unique character is revealed in moments of performance. P, admitted that answers are shaped by environments, highlighting the role of situation. Q, even in similar conditions, people's responses may differ depending on their personal beliefs. R, replied personal beliefs can lead to different responses, even when the situations are alike. S, showing that altered perspectives can arise under different conditions. T, stated that in a settings behave combination of various traits from multiple personalities. However U, demonstrates that their attitude significantly impacts how they engage with others.

iii. Classroom Management

Another theme that emerged from the analysis was classroom management. Silence is considered effective for classroom management, and teachers also use group activities as a strategy for managing the classroom. The university teachers mentioned this in their responses

1) Silence is Effective for Classroom Management

A and B, both expressed doubt about the significance of silence. However, C, highlighted the positive effect of silence on raised focus and attentiveness through active listening. D, emphasizes that silence facilitates easier comprehension. E, states same concept that silence is essential for achieving deep understanding. In gap, F, disagreed and stated that silence in the classroom could lead to passivity among students. G, opposed this argument and stated that silence allowed teachers to convey concepts more effectively. On the other hand, H, emphasizes that the effectiveness of silence depends on the teacher's skill in utilizing it. I, uncertainty over the importance of silence. J, demonstrated silence enhance concentration and improve attentive listening. K, Showed that remaining silent can sharpen attention and deepen listening skills. L, highlights that quietness makes understanding simpler. M, true comprehension can only emerge in the presence of silence. N, arguing that a quiet classroom might result in students becoming passive. O, arguing that maintaining silence enabled teachers to explain ideas more clearly. P, concern regarding silence's importance. Q, emphasized the role of silence in boosting focus and encouraging active listening. R, Explained the benefits of silence in increasing mental clarity and fostering engaged listening. S, replied that being silent helps in grasping things more easily. T, silence creates the space needed for profound insight to take root. Lastly U, saying that too much silence in class could make students disengaged and inactive.

2) Group Activity is useful in classroom management

A, stated that group activities cover individual capabilities, while B, gave the idea that its effectiveness depended on various factors. C, added support by highlighting its efficacy in lab work management and active learning scenarios. D and E, take a part by stating that group activities can be beneficial when carried out through theme games or drama and role play. F, expanded on this by supporting for brainstorming activities within groups. However, G, disagreed and argued that team-based assignments might hide weaker students within the group. H, pointed out that group activities might not be suitable for individual learning. I, mentioned that group tasks often overshadow individual skills. J, suggested that how well it worked was influenced by several different factors. K, demonstrating its effectiveness in managing laboratory tasks and promoting active learning. L, engaging in group activities through themed games or dramatic role-playing can enhance their effectiveness and benefits. M, group settings can significantly boost the positive outcomes of such activities. N, facilitated group brainstorming sessions to encourage idea generation and collaboration. O, opposed that group assignments could allow less capable students to blend in without contributing significantly. P, observed that participating in group activities can mask personal abilities. Q, indicated that its success was determined by a range of variables. R, showing its capability in streamlining work and enhancing engagement in active learning environments. S, group tasks become more impactful when they incorporate role play or are structured as themed games. T, group participation can be more beneficial when activities involve creative methods. Lastly, U, Provided support for collaborative thinking exercises within teams.

iv. Commitment and Responsibilities

Another theme from the analysis of the semi-structured interview transcripts was commitment and responsibility. This theme relates to effective performance, which is influenced by teachers' commitment. The respondents believe that being committed increases one's sense of responsibility. The university teachers mentioned this in their responses

1) Performance intervened the commitment

A, stated that it couldn't affect, to which B, disagrees and states that it does truly have an effect. C, supported B's viewpoint, emphasized a strong correlation between the two factors. Moving beyond a link, D, highlighted the importance of preplanned commitments fulfilled by teachers in relation to departmental responsibilities. E, also emphasized the significance of preparedness in meeting these commitments. F, adds another dimension, states that performance is always linked to the completion of tasks, creating a teacher's commitment. G, further improves that idea by stating that fulfilling commitments demonstrates effective performance. H, makes people aware of the role of self-monitoring in offering effective performance and states that teachers who engage in self-evaluation develop to perform better. I, claimed that it was incapable of having an effect. J, argues that it genuinely does make a difference. K, insists that it indeed has a real impact. L, highlighted a significant link between the two variables. M, emphasizing self-monitoring shows its effect on achieving strong performance, and teachers who reflect on their work are more likely to grow and become more effective. N, emphasizing that performance is consistently tied to task completion, which in turn fosters teacher dedication. O, commitments is a clear indicator of strong performance. P, said it had no influence on the matter. Q, stated that a close relationship between the two elements. R, crucial it is for teachers to follow through on their predetermined responsibilities within the department. S, highlighting the importance of self-monitoring helps individuals enhance their performance, and educators who regularly assess themselves tend to improve their teaching effectiveness. T, stated that performance is inherently connected to fulfilling tasks, thereby reinforcing a teacher's sense of responsibility. Finally, U, commitments reflects an individual's ability to perform effectively.

2) Commitment Increase your Responsibility

A, believed in firm commitment, emphasized its ongoing presence. B, acknowledged commitment but hinted at its varied. In comparison. C, linked commitment to workload, stated its variations based on work demands. D, viewed responsibility as associated to commitment, with the responsibility to fulfill tasks as committed. E, recited the same idea, focusing on the necessity of responsibility for fulfillment. F, connected commitment with professionalism. G, emphasized the levels of professionalism through accountable commitment. Similarly, H, differed, defining commitment and responsibility as various views. I, held a strong belief in unwavering dedication and highlighted its continuous role. J, recognized the dedication, though noted its inconsistency In contrast. K, commitment was tied to the workload and adjusted depending on the intensity of work demands. L, believed that responsibility is tied to commitment, meaning one must complete tasks they have agreed to. M, emphasized the importance of taking responsibility to achieve one's goals. N, demonstrated dedication through a professional approach. O, demonstrated a high standard of professionalism by consistently honoring responsibilities. P, supported faithful commitment and stressed its lasting significance. Q, admitted to the presence of commitment, yet pointed out how it differed in degree compared to others. R, commitment was connected to the workload and varied in response to changing work requirements. S, responsibility as a reflection of one's commitment, requiring the completion of duties as promised. T, highlighting that fulfillment depends on accepting responsibility. Finally, U, Aligned strong commitment with a high standard of professionalism.

All participants in the conversation agreed that teachers should offer solutions to improve the learning process and support the use of a variety of teaching pedagogies in the classroom by highlighting their advantages. The discussion then delved into the strong association among teachers' mindset, reactions, circumstances, personality, and approach. They emphasized the importance of this relationship, acknowledged its possible limitations, and stressed the need to evaluate every detail of its involvement. Furthermore, a variety of perspectives and points of view emerged during the conversation, focusing on the effects of preparation, commitment, self-monitoring, and the relationship between commitment and responsibility on effective performance. The discussion also highlighted the value of teacher involvement in learning environments, emphasizing its critical role in supporting efficient instruction and promoting student achievement.

c. Effect of University Teachers Metacognitive Regularities on their Work Efficiency

Objective 3: To assess the effect of metacognitive regularities on work efficiency of university teachers.

Research Question 3. How do university teachers consider the effect of their metacognitive regularities on their work efficiency?

Themes Emerged

i. Metacognitive Regularities Effects on Work Efficiency

An additional key theme identified from the interviews was support from university teachers, emphasizing that monitoring its effects on teaching and evaluation helps integrate the information.

1) Enhance Teaching

A, argued that monitoring indeed enhanced teaching. While B, introduces degree by suggesting that monitoring may have positive effects but it's not always. Similarly, C, highlighted the potential negative impact of biased monitoring, indicating that its effectiveness depends on the fairness of its implementation. D, focused on its role in identifying potential issues. E and F, adds that monitoring influences the procedures of lecture delivery, contributing to its effectiveness. Conversely, G and H, oppose that monitoring serves merely as an indicator or a measuring parameter, respectively, suggesting a more limited role. I, stated that supervision genuinely improved instructional quality. J, degree suggests that monitoring can sometimes have beneficial effects, though this isn't always the case. K, emphasized that biased monitoring could have harmful consequences, noting that its success joints on how fairly it is applied. L, concentrated on its function of detecting possible problems. M, observing and tracking lecture delivery processes helps improve their overall effectiveness. N, belief that monitoring is solely a tool for measurement or indication, implying it has only a restricted function. O, indication that monitoring functions just as a gauge or metric, thus assigning it a narrow scope. P, claimed that overseeing teachers' work effectively strengthened their teaching performance. Q, implies that while monitoring might lead to positive outcomes, its impact isn't consistently favorable. R, replied that the negative effects of biased monitoring may arise if it's not implemented equitably, as its effectiveness relies on impartial execution. S, emphasized its purpose in spotting potential concerns. T, monitoring plays a role in shaping how lectures are delivered, thereby enhancing their impact. Finally, U, argue against the view that monitoring merely reflects or measures performance, as this interpretation limits its full significance.

2) Integrates the Information

A, claimed the identification of a superior option among the choices. B, through participation. C, added depth to the conversation by emphasizing the importance of evaluative clauses in determining the best choice. D, introduced the role of computers in facilitating evaluation processes and obtaining acknowledgment. E, mentioned the significance of informative data in evaluations, supported by Participant F, emphasized the presentation of results. G, expanded the idea by highlighting the practical integration of information during evaluations. H, contributed by mentioning the diversity of evaluation outcomes in report cards. I, stated that the best option had been identified from the available choices. J, by taking part. K, highlighting how evaluative clauses play a crucial role in identifying the most suitable option. L, explored how computers assist in streamlining evaluation procedures and gaining recognition. M, the importance of using informative data in evaluations was highlighted, with support from the participant. N, placed a strong focus on how the results were presented. O, emphasized the real-world application of information by incorporating it effectively during assessments. P, stated that one option stood out as the most favorable among the alternatives. Q, via active involvement. R, contributed the significance of evaluative clauses in making the optimal decision. S, discussed the use of computers in enhancing assessment methods and securing acknowledgment. T, participant emphasized the value of relying on meaningful data when conducting evaluations. Lastly, U, highlighted the importance of presenting the results clearly.

ii. Work Efficiency Determination

Establishing a vision for work efficiency at the university level requires a well-organized university environment and acknowledges that teachers influence students' overall well-being.

1) University Environment Remain Well Organized by Teachers Work

A, acknowledges this diversity and states that texts do not always connect. B, agreed, claiming the system of teachers for this gap. C, supported this view by emphasizing the self-rule teachers own within their classrooms. D, added the necessity of self-motivation for such diversity to flourish. However, E, argues that individual teachers alone cannot achieve that changes. F, presents a counterpoint and states that well-organized teachers can reduce negative influences within institutions, thereby G, disagrees, opposing that it's the responsibility of the administration to ensure such diversity through monitoring teaching staff. H, determined the discussion by highlighting the essential role of teachers in running educational institutions. I, stated that texts do not necessarily always align or relate to each other. J, yes, agreed that the responsibility for addressing this gap should fall on the education system and teachers. K, highlighting the autonomy teachers have within their own classrooms. L, the importance of self-motivation is essential for fostering such diversity to thrive. M, argued that individual teachers by themselves cannot bring about those changes. N, suggests that skilled and well-structured teachers have the ability to diminish adverse effects within educational institutions. O, belief that ensuring diversity among teaching staff falls under the administration's responsibility to monitor and manage. P, acknowledges the diversity present and emphasizes that texts may not always be directly linked or connected. Q, agree that closing this gap is a task for the educational system and its teachers. R, emphasizing the self-governing authority teachers possess in their classroom environments. S, stated that diversity to prosper, it is crucial to have self-motivation as a driving force. T, claim is made that only teachers working alone cannot accomplish those transformations. Finally, U, argued that effective and properly organized teachers can mitigate harmful influences present within school environments.

2) Teachers Affects Students Over all Well Being

A, acknowledges that this effect is dependent upon the level of engagement established by the teacher. B, agreed, emphasizing the crucial role of teachers in fostering student growth. C, adds to the conversation and highlights that teachers serve as modifiers for students, shaping their educational experience. D, agrees and claims that effective teaching is truly dynamic in this process. E, emphasized the importance of building positive and respectful relationships with students. F, expands this view and states that effective communication is key to successful teaching. G, pointed out the necessity of monitoring and guiding students in a timely manner to ensure their progress. H, defined the discussion by emphasizing the role of teachers in supporting students and boosting their academic potential. I, recognizes that this impact relies on the degree of student engagement fostered by the teacher. J, yes, highlighting the essential importance of teachers in promoting the development of students. K, teachers act as agents of change for students, influencing and shaping their learning journey. L, agreed and state that successful teaching is genuinely adaptable within this process. M, highlighted the significance of fostering positive and respectful connections with students. N, highlights that clear and efficient communication is essential for achieving success in teaching. O, emphasized the importance of supervising and directing students promptly to guarantee their development. P, understands that the extent of this effect is influenced by how effectively the teacher promotes engagement. Q, emphasizing the vital contribution of educators in nurturing student progress. R, Educators function as catalysts in students' development, molding their academic experiences. S, encourage that impactful instruction is genuinely flexible and evolving throughout this process. T, replied the value of developing supportive and courteous relationships with learners. Finally, U, emphasizes that mastering effective communication is fundamental to ensuring successful educational outcomes.

The results highlight that university teachers recognize the impact of their metacognitive practices—such as monitoring and evaluation—on their work efficiency, playing a crucial role in improving teaching effectiveness and integrating information. While some participants view monitoring as a valuable tool for enhancing lecture delivery and identifying teaching issues, others emphasize its limitations when not implemented fairly. Regarding work efficiency, a well-organized university environment is seen as partially dependent on teachers' efforts, though administrative support is also essential. Finally, the results clearly indicate that teachers significantly influence students' overall well-being, particularly through engagement, communication, and positive relationships.

6. Comparison of Result

The quantitative and qualitative data were based on two research questions. The first question investigated the metacognitive regularities of university teachers and aimed to determine their work efficiency. The study examined the effect of metacognitive regularities on the work efficiency of university teachers using a conceptual framework. The understanding of university teachers was obtained through a self-developed questionnaire and a semi-structured interview based on the theory of metacognitive regularities (Julie et al., 2021) and model of work efficiency (Siti et al., 2012) for study the practices of university teachers.

The results from the descriptive statistics revealed that the university teachers agreed on the metacognitive regularities. They agreed that planning enhances the work efficiency of respondents and is important for organizing the best university activities. They also agreed on monitoring as a key aspect of teachers' professional development and on evaluation as a means to promote institutional standards. Furthermore, the university teachers agreed on the importance of work efficiency, acknowledged the cognitive ability to utilize intellectual capacities in decision-making, and emphasized maintaining a versatile personality for effective work. They also agreed on the importance of classroom management, expressed a preference for designing classrooms that focus on students, and affirmed their commitment and responsibilities for effective university functioning. The inferential statistics reveal the positive influence of metacognitive regularities on the work efficiency of university teachers. The sub-variable concludes that planning, monitoring and evaluation also effects positively on work efficiency of university teachers.

The thematic analysis of the first research question indicated the importance of vision and mission in planning, the effective use of calendars, and regularities for task completion. It emphasized monitoring as essential for achieving goals, allocating resources, and maintaining productivity, despite its complexities. Evaluation was recognized as challenging yet critical, particularly in subject development and annual academic reviews. Participants also highlighted the significance of inspection and methodology in enhancing learning outcomes. For the second research question, the analysis underscored the importance of teachers providing solutions to improve the learning process and adopting diverse pedagogies, emphasizing their benefits. It explored the complex relationship between teachers' personality and approach, alongside their preparation, commitment, and monitoring. The participants acknowledged the significance of teacher involvement in fostering metacognitive regularity for work efficiency, student achievement, and maintaining a balance between commitment and responsibility to achieve successful outcomes.

The comparative results of the inferential statistics and thematic analysis indicate that metacognitive regularities; planning, monitoring, and evaluation are essential for enhancing work efficiency among university teachers. While the inferential statistics provide a direct and quantifiable link to work efficiency, the thematic analysis offers rich, context-driven insights into how these regularities are applied in teaching practices. Additionally, the thematic findings introduce dimensions related to teacher personality, involvement, and pedagogical approaches, which contribute to a broader, more holistic understanding of the factors that influence effective teaching and working outcomes. Altogether, these findings present a comprehensive view of university teachers' efficiency and effectiveness.

Table No. 4.11.
Quantitative and Qualitative Major Results

Quantitative Outcomes	Qualitative Outcomes
	Thematic analysis was performed on semi-structured interview sheet.
Descriptive Statistics Metacognitive regularity of university teachers. (mean score 3.97) Higher score	The first question was that are the metacognitive regularities of university teachers? The themes that emerged from the first interview question included, metacognitive regularity enhance teaching, integrates the information, planning comprise, planning effect on students' performance, planning improve course efficiency, planning complete commitment, planning schedule work in advance, monitoring organize best work, monitoring difficult to do, presentation help student work

	monitoring, achieving step by step commitment, class resources updated, provide factual support, evaluation enhance the quality of work, evaluation promotes subject inspection, evaluation found gap in teaching process and evaluation consider annual practice.
Work efficiency of university teachers (mean score 3.91) Higher score	The second interview question, what is the level of the work efficiency of university teachers? The themes that emerged from the second interview question included, University environment remain well organized by teachers work, teachers affects students over all well-being, cognitive ability helps in teaching practice, cognitive ability helps in class room management, personality affect the teaching, personality is the combination of attitude, silence is effective for classroom management, group activity is useful in classroom management, performance intervened the commitment, commitment increase your responsibility.
Inferential Out comes H_0 1= Effect of Metacognitive Regularities (Private and Public) rejected Significance difference at t(453)=5.512 where p.000 result is positively and significantly influence H_0 1a= Effect of Planning (Private and Public) rejected Significance difference at t(453)=9.183 where p.000 result is Positive effect on the work efficiency	The third interview question, How do university teachers consider the effect of their metacognitive regularities on their work efficiency? The themes that emerged from the third interview question included, university teachers consider the effect of their metacognitive regularities on their work efficiency such as monitoring and evaluation, play a crucial role in improving teaching effectiveness and information integration. While some participants view monitoring as a valuable tool for enhancing lecture delivery and identifying teaching issues, others stress its limitations if not implemented fairly. In terms of work efficiency, a well-organized

$H_0$1b= Effect of Monitoring (Private and Public)
rejected
Significance difference at
t(453)=7.96 where p.000
result is Positive effect on the work efficiency

$H_0$1c= Effect of Evaluation (Private and Public)
rejected
Significance difference at
t(453)=9.694 where p.000
result is Positive effect on the work efficiency

university environment is seen as partially dependent on teachers' efforts, though administrative support is also necessary.

a. Major Results

This study examined the effect of metacognitive regularities on the work efficiency of university teachers. The results showed that respondents agreed on the importance of planning for work efficiency, career development, and achieving targeted university outcomes. Monitoring was recognized as crucial for improving lecture delivery, tracking student progress, and fulfilling teaching commitments. Evaluation was acknowledged as essential for enhancing teaching quality, identifying learning gaps, and strengthening institutional standards. Overall, planning, monitoring, and evaluation were viewed as vital for professional growth and academic success. Furthermore, the results revealed that respondents' cognitive abilities, work efficiency, and intellectual skills were effectively applied to teaching and decision-making for students. Teachers' personalities and emotional control positively impacted relationships, discipline, and classroom behavior. Efficient classroom management strategies and teamwork were widely practiced with respect for regulations and resources. Teachers demonstrated strong commitment to responsibilities, student development, institutional activities, and maintaining academic standards. Additionally, statistical results showed that metacognitive regularities positively and significantly influence the work efficiency of university teachers. It was found that planning, as measured by metacognitive regularities, has a positive effect on the work efficiency of university teachers. Monitoring, as measured by metacognitive regularities, has a statistically significant effect on work efficiency, and evaluation also has a statistically significant positive effect on the work efficiency of university teachers. Moreover, the findings emphasized the critical understanding of the complexities and benefits of monitoring, evaluation, and annual reviews in enhancing learning and productivity. The discussion highlighted the challenges and importance of effective planning, monitoring, and evaluation in academic processes. The effect of metacognitive regularities on teachers should offer solutions and support in teaching methods by demonstrating their benefits. Participants also discussed how teachers' personalities influence their approach and relationships. They further highlighted the effect of commitment and responsibility on teaching effectiveness and student success.

b. Effects of the Findings

The findings of this study have several important implications for the professional development of university teachers and the broader academic environment.

Enhancement of Teaching Practices: The positive effect of metacognitive regularities planning, monitoring, and evaluation on work efficiency suggests that incorporating structured metacognitive regularities can improve lecture delivery, lesson planning, and assessment methods. Universities may benefit from training programs that strengthen teachers' metacognitive skills.

Professional Growth and Career Development: Since planning, monitoring, and evaluation were found essential for academic success, institutions may use these findings to design professional development programs that help teachers set goals, track progress, and assess outcomes effectively, contributing to career advancement and institutional excellence.

Improved Student Outcomes: Teachers' cognitive abilities, emotional control, and personality traits were shown to positively influence classroom management and student relationships. This implies that fostering both cognitive and emotional competencies in teachers can enhance student engagement, discipline, and overall learning outcomes.

Institutional Policy and Support: The study emphasizes the need for universities to provide systems that encourage regular planning, monitoring, and evaluation. Policies that support reflective practices, feedback mechanisms, and annual performance reviews can enhance productivity and teaching quality.

Decision-Making and Problem Solving: Teachers' application of cognitive skills in decision-making highlights the role of metacognitive awareness in solving classroom and administrative challenges. Institutions can encourage reflective practices that promote critical thinking and adaptive problem-solving.

Teamwork and Classroom Management: Findings indicate that effective classroom management and teamwork are facilitated by teachers' metacognitive awareness and emotional regulation. This suggests that collaborative teaching approaches and structured classroom management strategies should be emphasized in professional development initiatives.

Strategic Planning for Academic Excellence: The statistically significant relationship between metacognitive regularities and work efficiency underlines the importance of integrating planning, monitoring, and evaluation into everyday academic practices. Institutions can leverage this to align teaching practices with long-term educational goals and quality standards.

In summary, these findings suggest that universities should actively foster metacognitive skills, emotional intelligence, and structured teaching strategies among faculty to enhance efficiency, teaching effectiveness, and student success.

Chapter Five

Summary, Discussion, Findings, Conclusions and Recommendations

5.1. Summary

The research and understanding of regulations are fundamental for teachers' academic activity. Identifying motivating elements and implementing staff changes are essential for the university's competitiveness. The teacher's work efficiency may improve through incorporating metacognitive regularities. Teachers can set clear goals, collaborate efficiently, identify opportunities, and find multiple solutions for university students' educational problems. Metacognitive regularities are significant for teachers. It helps to manage workplace resources and institution requirements. It makes teaching-learning practices effective and manages lecturers successfully. This study investigates the effect of metacognitive regularities on university teachers' work efficiency. Specifically, the objectives are: To investigate the metacognitive regularities of university teachers, to determine the work efficiency of university teachers at the university and to assess the effect of metacognitive regularities on the work efficiency of university teachers. This includes explaining the specific effects of planning, monitoring, and evaluation on their work efficiency through Sub-objectives. The mix methods quantitative data got deep insight through the following research questions: 1) what are the metacognitive regularities of university teachers? 2) How can the work efficiency of university teachers at the university be determined? How do university teachers consider the effect of their metacognitive regularities on their work efficiency? These objectives and questions collectively investigate, how metacognitive regularity practices influence the work effectiveness of university teachers.

This study describes; the effectiveness of metacognitive regularities in university teachers working. The created conceptual framework gives a clear base to establish hypotheses and research questions, for improving work behavior in universities. The conceptual framework is based on the root theory of metacognition. The theory of mind known as metacognition examines mental processes in individuals, the current study was to make clear the mechanisms behind metacognitive regularities in university instructors' work efficiency. It is a belief that implies general mental processes connected to the work and metacognitive regularities refer to the processes that control people's cognitive operations and identify their perceptions. The conceptual framework is established through the model of metacognitive regularities (Julie et al., 2021) and the work efficiency model (Siti et al., 2012). It can be useful in monitoring development and potential. It can help advance the development of productive teamwork and provide multiple solutions to issues raised by learners. Metacognitive regulation model refers to becoming aware of efficiency as necessary for metacognition and it reflects in selected suitable resource management. It has a direct effect on performance outcomes. It has three sub-variables such as; planning, monitoring and evaluation. Work efficiency model helps to study understanding and focusing on tasks within an organization, keeping track of shared objectives and achieving desired results. It has four sub-variables such as: cognitive ability, personality, classroom management and commitment and responsibility. There are lack of metacognitive regularities research in the field of education sciences and it might be filled by this current study. Instructors are important assets for higher educational institutions, and their work reflections can be used to gauge workplace sustainability.

The explanatory study utilized pragmatism, combining epistemology and reality to investigate university teachers' work efficiency in pedagogical and organizational practices. This mix-methods research approach includes quantitative and qualitative data analysis. The effect of metacognitive regularities on work efficiency was described through descriptive correlational design. The total population of 23 Islamabad universities are 10464, of which 7743 were from public sector universities and 2721 were from Private sector universities. The proportionate stratified sampling techniques were used to select public and private sector university teachers for quantitative data. The sample size was 05% (524) of the total population, the 388 university teachers were selected from the public sector and 136 from private sector universities. Purposive sampling was used for qualitative data and it was approximately 04% of the quantitative sampling, the 15 university teachers were selected from the public sector and 06 from private sector universities. Quantitative data was collected by the questionnaire and it was measured by a five-point Likert scale. The questionnaire have two self-developed scales, the metacognitive regulatory assessment scale further has 3 sub-variables given by (Julie et al., 2021) and the work efficiency assessment scale more has the 4 sub-variables given by (Siti et al., 2012). Qualitative data collected by Semi-structured interviews. The delimitation of the study included universities in Islamabad with HEC recognition, the study scope was delimited to the investigation of a one-way effect, research exclusively based on university teachers at the universities of Islamabad. The instruments were verified by validity and reliability tests. The validity test was done by four educational experts. The 52 university teachers were selected for pilot test and 38 returned that was a 72% rate of return. The reliability of the teacher metacognitive regularities scale was 0.959, indicating that the scale is a reliable measure of teachers' metacognitive regularities and the reliability of the teacher work efficiency scale, consisting of 42 items and alpha=0.983, shows excellent internal consistency for the complete scale. The reliability test was analyzed by SPSS version 20th. After the data collection, the effect of metacognitive regularities on the work efficiency of university teachers' mean score and regression analysis were used. The conclusion and recommendations were made upon the analysis and findings.

5.2 Findings

The research findings are presented under this heading.

Objective No. 1. "To investigate the metacognitive regularities of university teachers."

1. *The university teachers in Islamabad generally agreed with (mean=3.97) metacognitive regularities as reflected in their responses.*
2. *Metacognitive regularities was an independent variable that included three sub-variables. Planning was the most prominent sub-variables with (mean=4.07) university teachers expressing agreement at the highest level compared to the other sub-variable.*
3. *The monitoring results (mean=3.96) indicate that university teachers generally agreed and representing a moderate level among the sub-variables of metacognitive regularities.*
4. *The evaluation indicates that university teachers agreed through this was the lowest level of agreement among the sub-variables of metacognitive regularities (mean=3.84).*

Objective No. 2. "To determine the work efficiency of university teachers."

1. *The university teachers generally agreed to their work efficiency (mean=3.91).*
2. *Work efficiency was the dependent variable and included four sub-variables. Among these, cognitive ability was noted as an area where university teachers agreed (mean=3.96) making it the second most highly rated sub-variable of work efficiency.*
3. *The personality aspect received the strongest agreement from university teachers and making it the most prominent sub-variable related to work efficiency (mean=3.99).*
4. *The classroom management aspect indicates that university teachers generally agreed (mean=3.77) through it was identified as the least effective among the sub-variables of the work efficiency.*
5. *The commitment and responsibilities of university teachers were found to agree (mean=3.91) and reflect a moderate level of work efficiency among the sub-variables assessed.*

Objective No. 3. "To assess the effect of metacognitive regularities on the work efficiency of university teachers."

1. *It was found that metacognitive regularities account for a significant portion of the variance in work efficiency with a strong relationship observed between the independent and dependent variables as indicated by the regression analysis (*$\boldsymbol{R^2} = \mathbf{0.831}$).
2. *The beta coefficient (β=0.415) indicates a positive relationship of metacognitive regularities.*
3. *The value of t=5.512 measured the statistical significance of the relationship between the independent and dependent variables at a significance level of one per cent and the findings was statistically significant.*

Objective No. 3a. "To assess the effects of planning on work efficiency of university teachers."

1. *It was found that planning accounts for a significant proportion of the variance in work efficiency with the regression (*$\boldsymbol{R^2} = \mathbf{0.756}$)*value indicating a strong relationship.*

2. *The beta coefficient (β=0.869) indicates a strongly positive relationship with planning.*

3. *The value of t=9.183 measured the statistical significance of the relationship between planning and work efficiency and indicated a significantly high.*

Objective No. 3b. "To assess the effects of monitoring on work efficiency of university teachers."

1. *It was found that monitoring for a significant part of the variance in work efficiency with the regression (*$\boldsymbol{R^2} = \mathbf{0.760}$) *value indicates a strong correlation.*
2. *The beta coefficient (β=0.872) indicates a strongly positive relationship with monitoring.*
3. *The value of t=7.967 measured the statistical significance of the relationship between monitoring and work efficiency and the finding was statistically significant.*

Objective No. 3c. "To assess the effects of evaluation on work efficiency of university teachers."

1. *It was found that evaluation was a significant part of the variance in work efficiency with a high regression ($R^2 = 0.706$) coefficient value indicates a strong relationship.*
2. *The beta coefficient (β=0.840) indicates a strongly positive relationship with evaluation.*
3. *The value of t=9.694 measured the statistical significance of the relationship between evaluation and work efficiency and indicated that the finding was statistically significant.*

a. Quantitative and Qualitative Comparative Findings

Metacognitive regularities were investigated in the practices of public and private sector university teachers. The effect of these regularities was observed in their cognitive ability, personality, classroom management, and sense of commitment and responsibility. The analysis highlighted the role of planning in achieving university tasks, along with the importance of evaluation and monitoring. Teachers' personalities and classroom management skills are essential for enhancing work efficiency and fostering student achievement. Maintaining a balance between responsibility and commitment remains key to successful educational outcomes. University teachers also acknowledged the significance of metacognitive regularities in improving their work efficiency.

5.3. Discussion

The study was established on three main objectives and five hypotheses (one main & three sub-hypotheses). The findings of the research provided valuable insights related to the effect of metacognitive regularities on work efficiency of university teachers. The first main objective of the study focused on investigating the metacognitive regularities in university teachers. The majority of the teachers were aware of metacognitive regularities in their practices. The study found that respondents agree on the importance of planning for work efficiency, including creating a conducive learning environment, and career development activities through planning and managing class time. University teachers also stated that overseeing their metacognitive regularities is essential for enhancing their delivery of lectures, ensuring professional development, and updating curriculum resources. They also agree that annual evaluation of courses, classroom observations, and teaching tasks is essential for improving teaching, detecting gaps, and addressing policy challenges. Therefore, Teachers are committed to evaluating themselves and their teaching tasks. Markeya (2016) views support that throughout their systematically review of the results, teachers' metacognitive teaching was highlighted as being essential to helping students develop metacognition. The range of performance results are associated with metacognition and those who are developing their cognitive capacities will most profit from metacognitive instruction.

The second major objective focused on determining the work efficiency of university teachers. It was found that most university teachers were aware of work efficiency for their successful teaching careers. The study found that teachers utilized cognitive abilities which are beneficial for pedagogy and classroom instruction. They use cognitive abilities in their work to manage tasks and make decisions. Teachers believe in strong relationships with students through strict discipline, and adapting to students' behavior. In the same context here, Hanna et al. (2019) study supports that the metacognitive-self helps in self-regulatory tasks. University teachers use teamwork strategies, respect university regulations, and focus on student motivation. They also maintain academic standards, engage students, and maintain a positive attitude. Teachers committed themselves to their subjects, participate in institutional activities, and uphold academic standards. They are committed to updating policies, evaluating classrooms, and respecting senior comments. Overall, teachers demonstrate effective work efficiency, commitment, and responsibility. Here previously, Ana and Mihaela (2018) explains in non-educational filed that the roles of managers can be characterized as centers of interaction, indicating that they are responsible for facilitating communication inside an institution. They completed the curriculum with responsibility, engaged students and participated in institutional activities. Teachers believed in maintaining academic standards, developed students' interpersonal abilities, they updated policies, evaluated classrooms, and respected senior comments and were part of university workgroups. It is effectively support by Roger and Vincent (2013) states work efficiency is the most essential element in success and responsibility since it creates possibility for positions in the future.

The third major objective assess the effect of metacognitive regularities on the work efficiency of university teachers. It explores the relationship between metacognitive regularities and work efficiency. First, the relationship of metacognitive regularity and work efficiency has discussed. A highly significant relationship was found between the strong explanatory powers of the independent variable (metacognitive regularities) on the dependent variable (work efficiency). It is supported by many studies such as; Harry (2004) states the institution can effectively accomplish sustainability by improving staff members' efficiency and developing team and individual capacity through a systematic and integrated approach. Therefore, Qiao (2019) supports that a novel method for assessing the quality of education provided by universities integrates contemporary strategies to evaluate and enhance institutional performance. Secondly, the relationship of sub-variable of metacognitive regularity with work efficiency of university teachers discussed. The sub-variable of planning was statistically significant, has a positive effect that was strong indication that planning affects work efficiency. These findings highlighted the significance of the purpose and vision in planning, and their function in education. In general, the idea is that finishing tasks ahead of schedule is a sensible and useful strategy regardless of opinions regarding its applicability in all circumstances. Furthermore, Indra (2023) supports that planning is the process of doing things through a series of related actions. Everyone must be ready through the planning of classes.

It was found that strong positive relationship between monitoring and work efficiency. The value of monitoring in work processes has a variety of perspectives on the significance of monitoring in a range of their related work areas with responsibility and resource allocation. Moreover, Xiaoyu et al. (2019) relates that the dynamic process of thought "self-monitored" itself by utilizing component knowledge and systematic analysis together with metacognition to monitor other processes.

It was found that evaluation has a statistically significant positive effect on the work efficiency of university teachers. The significance of evaluation in their work and underlined its difficult role in output. They both agreed that topic inspection and subject development are essential components of the learning process. In their academic course, they acknowledge their function as a thorough review of their procedures. The study supports by Eric and John (2012) that numerous distinct abilities and actions are reviewed in the evaluation, such as; planning and classroom management. The application of metacognitive regularities in the teaching-learning process is crucial to incorporating teachers into their work and beneficial for the institution and themselves.

5.4. Conclusion

The current study's findings result in the following conclusions.

The first objective focused on investigating the metacognitive regularities of university teachers. It was concluded that university teachers in Islamabad generally agreed on the presence of metacognitive regularities. Among the sub-variables of metacognitive regularities, planning emerged as the most prominent, reflecting a strong agreement among teachers on its importance. Monitoring held a moderate position indicating agreement but slightly less emphasis compared to planning. Finally, evaluation though still indicating agreement, suggesting it might be less prioritized or challenging in practice. These results highlight a varied but positive engagement with metacognitive practices among university teachers with a particular emphasis on planning.

The second objective to determine the work efficiency of university teachers were concluded the work efficiency of 453 university teachers in Islamabad, focusing on four sub-variables: cognitive ability, personality, classroom management, and commitment and responsibilities. The overall mean score for work efficiency, indicating that university teachers generally agreed on their efficiency. Moreover, Personality had the highest mean score suggesting strong agreement among university teachers regarding its contribution to work efficiency. Then cognitive ability ranked second highlighting its significant role in their work efficiency. Then Commitment and responsibilities scored moderately and reflecting agreement on its importance. Lastly, Classroom indicating it is an area where teachers felt comparatively less confident though still agreed on its importance.

The third objective assessed the effect of metacognitive regularities on the work efficiency of university teachers. It was concluded that metacognitive regularities significantly influence work efficiency. A strong positive relationship between metacognitive regularities and work efficiency was found.

5.4.1. Comparative Conclusion

The comparative conclusion highlights that metacognitive regularities, which integrate essential information, are crucial for enhancing teaching quality and making informed decisions. While monitoring can yield positive outcomes, its effectiveness depends on fairness and proper implementation. Overall, combining thorough evaluation methods with fair monitoring practices leads to more effective educational improvements. Planning for educational and organizational success underscores the importance of effective work management and monitoring in enhancing productivity and achieving objectives. The participants emphasized the significance of evaluation, despite its challenges. These insights reflect the complexities of monitoring and evaluation methodologies in education, emphasizing their role in fostering growth and development. Teachers play a pivotal role in promoting diversity, guiding student development, and shaping educational success through their autonomy, motivation, and effective engagement strategies. Their influence is essential in creating positive, inclusive learning environments that support overall well-being and academic growth. The essential role of university teachers is to enhance work efficiency and develop a deep understanding of their cognitive abilities and personalities, while emphasizing the importance of monitoring and responsibility. It became clear that teacher involvement is crucial for creating supportive classroom management that drives student achievement and effective instruction.

5.5. Recommendations for Teachers

The study revealed positive outcomes in its findings and conclusion;

however, to further enhance the work efficiency of university teachers, recommendations were provided for them.

1. *The study concludes that university teachers were familiar with metacognitive regulation practices. However, to enhance work efficiency, it is recommended that, if necessary, the university organize three-month short-term courses on metacognitive regulation for teachers, with a focus on their future career development.*
2. *Teachers undergo an annual evaluation process conducted by their universities. It is recommended that they prepare self-analysis reports to highlight their work efficiency. Writing these reports allows them to effectively showcase their contributions during the university's annual course evaluation.*
3. *Teachers can monitor their teaching practices to align with the university's future goals by staying informed about the latest curriculum for degree courses. This approach also enhances their work efficiency through regular monitoring of pedagogical methods. It is recommended to ensure effective planning in accordance with the updated academic curriculum.*
4. *Recorded teaching observations help fulfill commitments and responsibilities. It is recommended that teachers evaluate their work experience using a daily self-observation chart.*
5. *Managing evaluations, classroom activities, and organizational tasks can be challenging within a busy schedule. To address this, it is recommended to maintain a daily reflection manual as an informal record for evaluating daily activities.*

5.6. Recommendations for Future Researchers

The study's outcomes suggest that future researchers should take the following recommendations into account.

1. *This study shows the effect of metacognitive regularities on the work efficiency of university teachers. However, other branches of metacognition could also be explored in this context. Future researchers are encouraged to expand this investigation by*

conducting experimental studies on adaptive metacognition and its influence on the work efficiency of university teachers.

2. *The study of metacognitive regularities was initially conducted in the higher education sector. It is recommended to extend this analysis to the K–12 level. Future researchers are encouraged to investigate the effect of metacognitive regularities on the work efficiency of college teachers.*
3. *This study was conducted at Islamabad University, and its findings cannot be generalized to university teachers across Pakistan. It is recommended that similar research be conducted with university teachers from other cities in Pakistan.*

Biblography

Alejandro, P. A. (2015). *Metacognition: Fundaments, Applications, and Trends, A Profile of the Current State of Art, A Conceptual Model of the Metacognitive Activity* (pp.40). Springer, Heidelberg, New York, London, Library of Congress.

American Psychological Association. (2017). *Ethical Principles of Psychologists and Code of Conduct.* Retrieved from https://www.apa.org/ethics/code/

Bruce, L. B. (2001). *Qualitative Research Methods for the Social Sciences (4th ed.).* Allyn and Bacon, California, Pearson Education Company.

Burke, R. J., & Larry, C. (2014). *Educational Research Quantitative Qualitative and Mixed Approaches* (5thed.). Sage Publication ltd, London, United Kingdom, Library of Congress Cataloging-in-Publication Data. Retrieved from https://www.researchgate.net/publication/264274753

Creswell, J. W. (2009). *Research Design, Qualitative, Quantitative and Mixed Methods Approaches (3rd ed.).* University of Nebraska Lincoln, Sage , London.

Creswell, J. W. (2015). *Educational Research: Planning, Conducting, and Evaluating Quantitative (5th ed.*). Nebraska Lincoln, Pearson.

Creswell, J. W. (2018). *Educational research: Planning, conducting, and evaluating Quantitative and qualitative Research. 6th Edition (p. 206, 209).* Upper Saddle River, NJ: Prentice Hall

Donna, W., & Marcus, C. (2016). *Teaching Students to Drive their Brains*: Metacognitive Strategies, Activities, and Lesson Ideas. Alexandria Virginia: USA: ASCD. Retrieved from https://files.ascd.org/staticfiles/ascd/pdf

Edmonds, W. A., & Thomas, D. K. (2017*). An Applied Guide to Research Designs, Quantitative, Qualitative and Mixed Methods* (2nded.). Sage, United Kingdom.

Flavell, J. H., & Wellman, H. M. (1975). *Metamemory. Eric, University of Minnesota.* Retrieved from https://files.eric.ed.gov/fulltext/ED115405.pdf

Gay, L. R., Mills, G. E., & Airasian, P. (2012). *Educational Research Competencies for Analysis and Applications* (10th ed.). Merrill, Pearson.

Harry, T. (2004). *Educational Leadership Personal Growth for Professional Development*. London, Sage Publication.

Hope, J. H. (2001). *Metacognition in Learning and Instruction*. Kluwer, Academic Publisher.

James, C., & Peter, G. (1993). *Conceptualizing Reflection in Teacher Development.* Tylor & Francis, London, Washington. Retrieved from https://www.amazon.com/Conceptualising-Reflection-Teacher-Development-1993

Janet, M., & Arthur, P. S. (1994). *Metacognition; Knowing About knowing*. USA. Retrieved from https://books.google.com.pk/books?hl=en&lr=&id

Jean, P. (1936). *Origins of intelligence in the child*. London: Routledge & Kegan Paul. Retrieve from https://archive.org/details/in.ernet.dli.2015.458564/page/n11/mode/2up

John, B., Sharon, D., David, B., Karen, H., & Kelly, L. B. (2005). *Preparing Teachers for a Changing World, What Teachers Should Learn and Be Able to Do.* Wiley, Jossey-Bass, Washington. Retrieved from https://news.cehd.umn.edu/wp-content/uploads/2010/01/chapter-two.pdf

Jonathan, W. S., Michael, D. M., Benjamin, B., & Piotr, W. (2015). *APA Handbook of Personality and Social Psychology,* (Vol. 1), Ch3. Retrieved from http://search.worldcat.org/title/APA-handbook-of-personality-and-social psychology/oclc/885413705

Judit, O. G. (2016). *Training Programme for Teachers' Professional Development on Problem Solving.* Sense Publishers, Boston, Taipei, Library of Congress, Hungary, Ch4, 37-93. doi: 10.1007/978-94-6300-711-5_4

Julie, P. (2001). *SPSS Survival Manual.* Mc Graw-Hill, Canberra, Austrailia, Oxford Press.

Louis, C., Lawrence, M., & Keith, M. (2018). *Research Methods in Education,* (8th ed.). Routledge, Taylor & Francis group, London and New York. Retrieved from https://www.routledge.com/Research-Methods-in-Education/

Magdalene, L., & Deborah, L. B. (1999). *Teaching as the Learning Profession: Handbook of Policy and Practice.* San Francisco: Jossey-Bass. Retrieved from https://public.websites.umich.edu/~mlampert/LampertBallAligninTeacher.pdf

Margaret, M. (2022). *Metacognition, Thinking about Thinking* (4th ed.). Studocu. Retrieved from https://www.studocu.com/ph/document/xavier

Nilson, L. (2013). *Creating Self-Regulated Learners. Strategies to Strengthen Students' Self-Awareness and Learning Skills* (1st ed.). Stylus Publishing, Routledge, New York.

Norman, B. (2003). *Analyzing Quantitative Data: from description to Explanation.* London, Thousand Oaks, Sage Publication.

Rachel, E. McC. (2014). *Thinking about Thinking in Study Groups: Studying Engineering Students' Use of Metacognition in Naturalistic Settings.* Research Gate. Virginia Polytechnic Institute and State University, Tennessee. doi: 10.13140/RG.2.2.27077.14567

Roger, A., & Vincent, A. (2013). *International Handbook of Metacognition and Learning Technologies.* London: Springer. Retrieved from https://link.springer.com/book/10.1007/978-1-4419-5546-3

Schuyler, W. H. (2012). *Reading Statistics and Research,* (6th ed.). Pearson, Allyn & Bacon, Boston.

Schuyler, W. H. (2012). *Reading Statistics and Research,* (6th ed.). Pearson, Allyn & Bacon, Boston.

Whitebread, D., & Pino-Pasternak, D. (2013). *Interpersonal Regulation of Learning and Motivation: Methodological Advances,* Video Analysis of Self-Regulated Learning in Social and Naturalistic Contexts (pp.14-44). Oxon, Routlegde. doi: https://doi.org/10.4324/9780203117736

William, M. T., James, P. D., & Kanika, A. (2016). *Research Methods, The essential Knowledge Base.* Cengage Learning, Austrailia.

References

Aamir, A. C., & Sohail, Z. (2006). Antecedents and Consequences of Organizational Commitment among Pakistani University Teachers. *Applied H.R.M. Research, 11*(1), 39-64. Retrieved from https: 56d6cacb08aee73df6c2e48b

Afzal, S. M., & Md, A. K. (2021). Teaching and Learning Process to Enhance Teaching Effectiveness: a Literature Review. *International Journal of Humanities and Innovation, 4*(1), 1-4. Retrieved from https: https://files.eric.ed.gov/fulltext/ED610428.pdf

Aguirre, S. L. (2016). Evaluación De Unapropuesta Para Eldesarrollo De Laescrituraen Estudiantes Universitarios Apartir De Habilidades De Metacognición. Logos. *Revista de Lingüística, Filosofíay Literatura, 26*(2), 181–196. doi:dx.doi.org/10.15443/RL26015

Ana, G. C., & Mihaela, P. (2018). Management of Educational Efficiency and Efficiency. *Holistica, 9*(3), 89-96. doi: 10.2478/hjbpa-2018-0025

Anatoly, V. K., Elena, V. K., & Anna, V. C. (2023). Patterns of Development of Metacognitive Regulation of Information Activity in the Process of Professional Training. *ResearchGate*, 1032744.

Anna, V. (2021). Education and Management Practices. *Centre for Economic Performance* (Vol. 1767). Retrieved from https://cep.lse.ac.uk/pubs/download/dp1767.pdf

Anne, P., Tara, K., & Linda, D. H. (2019). Does Teaching Experience Increase Teacher Effectiveness? A Review of US Research. *Journal of Professional Capital and Community, 4*(3). doi: 10.1108/JPCC-12-2018-0032

Angelina, M., Thomas, D., & Brian, D. (2022). Metacognitive Knowledge, Regulation, and Study Habits. *Psychology, 13*(12). doi: 10.4236/psych.2022.1312112

Andreas, S. (2018). The Future of Education and Skills Education 2030. Retrieved from https://www.oecd.org/education/2030-project/contact/

Arifa, A., Muhammad, A. H., & Nazrul, I. (2019). Impact of Work Life Balance on Organizational Commitment of University Teachers: Evidence from Jashore University of Science and Technology. *International Journal of Scientific Research and Management, 7*(4), 1073-1079. Retrieved from https://.ijsrm.in ISSN (e): 2321-3418

Athanasios, D., Eleni, M., & Charalabos, S. (2023). Meta-Learning: A Nine-Layer Model Based on Metacognition and Smart Technologies. *Sustainability, 15*(2).doi:10.3390/su15021668

Baker, P. A., Lamoureux, K., Ken, W. McC., & Donald, J. T. (2013). The Amphitheatre Model for Talent Development: Recognizing and Nurturing the Gifts of our Lost Prizes. *International Journal for Talent Development and Creativity, 1*(1), 99-112. Retrieved from https://www.ijtdc.net/images/pdf/IJTDC_11_2013_Web.pdf

Bandura, A. (1989). Human Agency in Social Cognitive Theory. *American Psychologist, 44(9),* 1175. Retrieved from https://www.simplypsychology.org/social-cognitive-theory.html

Borg, S. (1998). Teachers' pedagogical systems and grammar teaching: A qualitative study. *TESOL Quarterly, 32(1),* 9-38. doi.org/10.2307/3587900

British Educational Research Association. (2018). *Ethical Guidelines for Educational Research.* Retrieved from https://www.bera.ac.uk/publication/ethicalguidelines-for-educational-research-2018

Brown, G. T. L. (2004). Teachers' Conceptions of Assessment: Implications for Policy and Professional Development. *Assessment in Education, 11(3).* 301-18. doi.org/10.1080/0969594042000304609

Brown, & Mclntyre, D. (1993). Making Sense of Teaching. *Biochemical Education, Open University Press, Buckingham and Philadelphia 21 (2),* 113. Retrieved from https://iubmb.onlinelibrary.wiley.com/doi/pdf/10.1016/0307-4412%2893%2990076-C

Bruce, D. Mc. D., Jeremy, L. H., Janine, O'F., & Sandra, V. T. (2022). The Future of Public Administration Research: Aneditor's Perspective. Wiley. *Public Administration, 100*(7). doi:10.1111/padm.12829

Charles, A. (2018). Types of Educational Planning/Reasons for Planning Education. *Research Gate*. Retrieved from https://www.researchgate.net/publication/338774755

Complete Dissertation. (n.d.). *By Statistics Solutions, Expert Guidance Every Step of the Way.* Retrieved from https https://trymata.com/blog/sampling-methods-in-qualitative-research/

Damien, S. F., Bert, B., & Wouter, V. D. B. (2021). Metacognition: Ideas and Insights from Neuro- and Educational Sciences. *Nature Partner Journals, Science of Learning 6*(1), 13. doi:10.1038/s41539-021-00089-5

Daniel, M., & Christian, B. (2020). Metacognition and Self-Regulation Learning: Evidence Review. *Education Endowment Foundation* (Vol. 1), 4-21. Retrieved from https://d2tic4wvo1iusb.cloudfront.net/

Daniela, P., Mario, M. F., Tania, A., Ceres, M., Teresa, B., & Pedro, D. (2021). Metacognitive Challenges to Support Self-Reflection of Students in Online Software Engineering Education. IEEE Xplore (2021). *Proceedings of the 4TH International Conference of the Portuguese Society for Engineering Education (CISPEE), June 21-23, 2021.* Lisbon, Portugal: IEEE. doi: 10.1109/CISPEE47794.2021.9507230

David, P. W. (1998). Coalitionary Mate Guarding by Male Chimpanzees at Ngogo, Kibale National Park, Uganda. *Behavioral Ecology and Sociobiology* (Vol.44), 43-55. Retrieved from https://link.springer.com/article/10.1007/s002650050513

Di, L. (2021) EFL Teachers' Optimism and Commitment and Their Contribution to Students' Academic Success. *Educational Psychology,* (Vol. 12). Retrieved from https://doi.org/10.3389/fpsyg.2021.752759

Dongho, K., & Cheolil, L. (2017). Promoting Socially Shared Metacognitive Regulation in Collaborative Project-Based Learning: a Framework for the Design of Structured Guidance. *CrossRef, 1*(6), 194-211. Retrieved from https://doi.org/10.1080/13562517.2017.1379484

Elena, S. (2014). From the Teachers Professional Ethics to the Personal Professional Responsibility. *Acta Didactica Napocensia, 7*(4). Retrieved from https://files.eric.ed.gov/fulltext/EJ1053254

Emmy, V., Sjef, S., & Theo, B. (2017). Successful Learning: Balancing Self–Regulation with Instructional Planning. *Teaching in Higher Education, 23*(1), 1-16. doi:10.1080/13562517.2017.1414784

Eric, S. T., & John, H. T. (2012). Can Teacher Evaluation Improve Teaching? Evidence of Systematic Growth in the Effectiveness of Midcareer Teachers. *Education Next, 12*(4). Retrieved from http:/ www.educationnext.org/can-teacher-evaluation /improve-teaching/

Farhat, A. K., & Suhail, A. K. (2013). Metacognitive Reading Strategies in Relationship with Scholastic Achievement in Science of IX Standard Students of English Medium Schools in Aurangabad city. *Journal of Educational Studies Trend & Practices, 3*(1), 119-129. Doi:10.52634/mier/2013/v3/i1/1564

Fiona, B. J. (2003). The Benefits and Value of Metacognitive Awareness in the Workplace: Developing the Thinking Performer, *CiteSeerX*, (Vol. 1). Retrieved from https://citeseerx.ist.psu.edu/

Flavell, J. H. (1979) Metacognition and Cognitive Monitoring: A new area of Cognitive-developmental inquiry. *American Psychologist, 34*(10). 906–11. Doi: https://psycnet.apa.org/doi/10.1037/0003-066X.34.10.906

Gage, N. L. (1964). Toward a Cognitive Theory of Teaching. Sage journals, 66(5). doi: org/10.1177/016146816406500

Gloria, D. A. (2005). Improving Teaching: Enhancing Ways of Being University Teachers. *Higher Education Research & Development, 24*(4), 361-372. doi:10.1080/07294360500284771

Georgia, S., & Maria, H. M. (2017). Metacognitive Knowledge and Metacognitive Regulation in Self-Regulatory Learning Style, and in Its Effects on Performance Expectation and Subsequent Performance Across Diverse School Subjects. *Scientific Research An Academic Publisher, 8*(12), 2152-7199. doi: 10.4236/psych.2017.812125

Gylmira, S., Altynay, T., Sholpan, K., Aitzhanova, R., & Marat, B. (2016). Monitoring of the Educational Process during the Pedagogical Practical Training in School. *International Journal of Environmental and Science Education. 11*(10), 3532-3547. Retrieved from https://files.eric.ed.gov/fulltext/EJ1114741.pdf

Harris, D. N., & Rutledge, S. A. (2007). Models and Predictors of Teacher Effectiveness: A Review of the Literature with Lessons from (and for) Other Occupations. *Madison, WI: Teacher Quality Research.* Retrieved from http://www.teacherqualityresearch.org/models.pdf

Hanna, B., Roman, K., Pawel, K., & Rex, W. (2019). The Metacognitive Self: The Role of Motivation and An Updated Measurement Tool. *Economics and Sociology, 12*(1), 208-232. doi:10.14254/2071-789X.2019/12-1/12

Houichi, A., & Sarnou, D. (2020). Cognitive Load Theory and its Relation to Instructional Design: Perspectives of Some Algerian University Teachers of English. *Arab World English journal, 11*(4), 110-127.

doi: 10.24093/awej/vol11no4.8

Ilona, S., & Sarka, H. (2014). Responsibilities and Competences of a University Teacher. *Social and Behavioral Sciences,* (Vol. 159), 437-441. doi:10.1016/j.sbspro.2014.12.403

Indra, P. (2023). Strategic Planning in Improving Organizational Performance: Analysis of Practice in Schools. *International Journal of Social Science and Human Research, 6*(7), 4359-4365. doi:10.47191/ijsshr/v6-i7-63

Jalal, H. (2016). Improving Employee Productivity through Work Engagement: Empirical Evidence from Higher Education Sector. *Growing Science,* (Vol. 6), 61-70. Retrieved from https://www.growingscience.com/msl/Vol6/msl_2015_114.pdf

Jared, A. L., Brain, P. T, & Lea, M. K. (2014). Nonparametric Statistics for Applied Research. *Springer, Newyork, London.* doi:10.1007/978-1-4614-9041-8

Jeffrey, M., Pamela, H. K., & Donetta, J. C. (2009). The Impact of a Social Cognitive Theory-based Intervention on Physical Education Teacher Self-Efficacy. *Professional Development in Education, Research Gate, 35*(4), 511-529. doi: 10.1080/19415250902781814

Jeltsen, P., Free, D. B., Veronique, R. R., Ankelien, K., Tine, B., & Koen, L. (2014). The Role of Teachers' Self-Regulatory Capacities in the Implementation of self-Regulated Learning Practices. Procedia. *Social and Behavioral Sciences, Elsevier,* (Vol. 116), 1963-1970. doi:10.1016/j.sbspro.2014.01.504

Jill, J., Maria, P., & Emmanuel, T. (2017). Efficiency in Education. *Journal of the Operational Research Society,* (Vol. 68), 331 - 338. Doi: 10.1057/s41274-016-0109-z

John, H. F. (1971). Stage-Related Properties of Cognitive Development. *Cognitive Psychology, Elsevier, 2(4),* 421- 453. Retrieve from https://doi.org/10.1016/0010-0285(71)90025-9

John, W. C., & Mary, L. D. (2019). Principals' Perceptions of Teacher Evaluation Reform from Structural and Human Resource Perspectives. *Journal of Educational Supervision, Empirical Research,* ERIC, *2*(1), 58-77. Retrieved from https://files.eric.ed.gov/fulltext/EJ1250308.pdf

Julia, C., Rebecca, J. J., & Jonathan, P. (2021). Defining Self-Awareness in the Context of Adult Development: A Systematic Literature Review. *Journal of Management Education, Sage, 46*(1), 1-38. Retrieved from https://doi.org/10.1177/1052562921990065

Julie, D. S., Amanda, J. S., & John, D. (2021). Fostering Meta Cognition to Support Student Leaning and Performance. *CBE-Life Sciences Education, 20*(3), 1-7. doi: 10.1187/cbe.20-12-0289

Katherine, A. B., & Kimberly, N. H. (2017). The liker Scale What It Is and How to Use It. *Military Operation Research Society, Phalanx, Jstor, 50(2),* 32-39. Retrieved from https://www.jstor.org/stable/26296382

Katherine, A. R., & John, D. (2013). Retrieval-Monitoring-Feedback (RMF) Technique for Producing Efficient and Durable Student Learning. *Springer, New York,* (Vol. 28), 67-78. doi:10.1007/978-1-4419-5546-3_5

Katherine, N. (1996). Language in Cognitive Development, Emergence of the Mediate Mind. *Journal of Evolution and Development Hybrid Mind, Research Gate, New York,* (Vol.3), 60-88. doi: 10.1017/CBO9781139174619

Khoirotun, N., Aslamiyah, A., Noorhafizah, N., & Novitawati, N. (2025). Teachers Work Effectiveness. *Educational Technologies, 2(2),* 291-295. doi.org/10.2991/978-2-38476-374-0_25

Killion, J., & Todnem, G. (1991). A Process for Personal Theory Building. *Educational Leadership, Eric* (Vol. 14). 14-17. Retrieved from https://eric.ed.gov/?id=EJ422847

Kinzie, J. (2019). Taking Stock of Initiatives to Improve Learning Quality in American Higher Education through Assessment. *Higher Education Policy, 32*(1). doi:10.1057/s41307-019-00148-y

Koster, B., & Dengerink, J.J. (2008). Professional Standards for Teacher Educators: How to Deal with Complexity Ownership and Function Experience from the Netherlands. *European Journal of Teacher Education, 31 (2),* 135-149. Retrieved from https://www.sciepub.com/reference/61976

Laila, M., & Fatima, B. (2020). Exploring Bem's Self-Perception Theory in an Educational Context. *Encyclopedia, 24*(58). Retrieved from https://doi.org/10.6092/issn.1825-8670/9891

Lalu, & Eny (2021). The Effect of Vigor, Dedication and Absorption on the Employee Performance of PT Garuda Indonesia Cargo. European Journal of Business and Management Research, 6(4). doi:10.24018/ejbmr.2021.6.4.1006

Lisa, B., & Robert, M. K. (2020). Smart Teachers, Successful Students? A Systematic Review of the Literature on Teachers' Cognitive Abilities and Teacher effectiveness. *Elsevier,* (Vol.30). Retrieved from https://doi.org/10.1016/j.edurev.2020.100312

Lisa, M. S., & Richard, A. C. (2020). Consistency, Not Speed: Temporal Regularity as a Metacognitive Cue. Psychological Research, *Springer, 84*(3), 88-98. doi:10.1007/s00426-018-0973-z

Marcel, V. J. V., Bernadette, V. H. W., & Peter, A. (2006). Metacognition and Learning: Conceptual and Methodological Considerations. *Metacognition Learning, 1*(1), 3-14. doi:10.1007/s11409-006-6893-0

Markeya, S. P. (2016). Fostering Metacognition in K-12 Classrooms: Recommendations for Practice. *A Student-Led Journal,* (Vol. 31). doi: 10.13014/K21Z429D

Marie, G., Sylvie, W., & Thierry, M. (2015). Beyond the Experience: Detection of Metamemorial Regularities. Consciousness and Cognition, *Elsevier,* (Vol. 33), 16-23. Retrieved from https://doi.org/10.1016/j.concog.2014.11.009

Martin, D., Oliver, D., & Markus, D. (2018). University Instructors' Achievement Goals for Teaching. *Journal of Educational Psychology, 111*(1), 131-148. doi:10.1037/edu0000271

Maziahtusima, I., Idayat, O., & Sakinah, A. (2025). The Role of Metacognitive Strategies in Enhancing Learning Outcomes and Educational Efficiency: A Systematic Review of Quantitative, Qualitative and Mixed-Method Studies. *International Journal of Academic Research in Business and Social Sciences, 15(4),* 81-101. doi:10.6007/IJARBSS/v15-i4/24964

Mehrdad, N. (2016). A Brief History Metacognition and Principles of Metacognitive Instruction in learning. *Journal of Humanities, Arts, Medicine and Science, 2*(2), 61-64. Retrieved from https://www.researchgate.net/publication/305996176

Meyer, J. P., & Allen, N. J. (1997). Commitment in the workplace. Theory, research and application. Thousand Oaks: Sage.

Mikhail, O., & Leona, P. (2020). The Role of Metacognition in Recognition of the Content of Statistical Learning. Psychonomic Bulletin & Review, Springer, (Vol. 28), 333-340. doi:10.3758/s13423-020-01800-0

Ming, M. C., & Sze, W. K. (2009). From Metacognition to Social Metacognition: Similarities, Differences, and Learning. *Journal of Education Research, 3*(4), 1-19. Retrieved from https://www.researchgate.net/publication/291046856

Muhammad, E. M., Samina, N., Basharat, N., & Rizwan, Q. D. (2010). Job Satisfaction and Organizational Commitment of University Teachers in Public

Sector of Pakistan. *International Journal of Business and Management, 5*(6). doi:10.5539/ijbm.v5n6p17

Nikolaos, K., Eleni, S., & Maria, K. (2023). Impact of Teachers' Demographics on Total Quality Management Parameters the Case of Primary Education. *Educational Sciences, 13*(7). Retrieved from https://doi.org/10.3390/educsci13070679

Nelson, N. (2018). Factors that Impact the Evaluation of Teacher Evaluation of Teacher Effectiveness for Reappointment, Tenure and Promotion. University of Bridgeport. Retrieved from https://files.eric.ed.gov/fulltext/EJ1209441.pdf

Nelson, T., & Narens, L. (1990). Metamemory: A Theoretical Framework and New Findings. Psychology of Learning and Motivation, Elsevier, 26, 125–173. doi.org/10.1016/S0079-7421(08)60053-5

Nosratinia, M., & Adibifar, S. (2014). The Effect of Teaching Metacognitive Strategies on Field-dependent and Independent Learners' Writing. Procedia. *Social and Behavioral Sciences,* (Vol. 98), 1390–1399. doi:10.1016/j.sbspro.2014.03.557.

Organization for Economic Co-operation and Development. (2019). OECD Future of Education and Skills 2030. Retrieved from https://www.oecd.org/content/dam/oecd/en/about/projects/edu/education-/Skills_for_2030

Paris, S., & Winograd, P. (1990). How Metacognition Can Promote Academic Learning and Instruction. *Educational Psychoogy, Tylor & Francis.*

Paul, R. P. (2010). The Role of Metacognitive Knowledge in Learning, Teaching and Assessing. *Theory in to Practices, Tylor & Francis, 41(4),* 219-225. doi: org/10.1207/s15430421tip4104_3

Paul, R. P. (1995). Understanding Self-regulated Learning. *Autumn, Wiley, 1995(63),* 3-12. doi:org/10.1002/tl.37219956304

Petros, G. (2004). From the General to the Situated: Three Decades of Metacognition. *International Journal of Science Education, 26*(3), 365-383. Doi:10.1080/0950069032000119401

Qiao, Y. P. (2019). Research on the Management of the Performance of College Ideological and Political Work Based on Personal Competency, Education and Humanities Research, (Vol. 281). *Proceedings of the 4th International Conference on Social Science and Contemporary Humanity Development,* 2018.

Randy, M. I., & Frank, F. (2006). Metacognitive Knowledge Monitoring and Self-Regulated Learning: Academic Success and Reflections on Learning. *Journal of the Scholarship of Teaching and Learning, 6*(1), 39-55. Retrieved from https://www.researchgate.net/publication/228626197

Rasha, M. A. (2020). Metacognitive Awareness and Academic Motivation and Their Impact on Academic Achievement of Ajman University students. *Heliyon, 6*(9). doi:10.1016/j.heliyon.2020.e04192

Rasmussen, J. (1985). The Role of Hierarchical Knowledge Representation in Decision and System Management IEEE Transaction on Systems, Man and Cybernetics, SMC, (Vol. 15). 234-243. Retrieved from https://www.scirp.org/reference/referencespapers?referenceid=2064660

Richard, J. S., (1973). What is the Basic Teaching Skill?. *Journal of Teacher Education, ResearchGate, 24(2),* 144-151. doi:10.1177/002248717302400213

Roberta, H., & Alison, T. (2015). Validity and Reliability in Quantitative Studies. *Research Made Simple, Cross Mark, Ontario Canada, 18(3)*. doi: 10.1136/eb-2015-102129

Roger, R. G., Adán, G. S., & Danilza, L. A. (2020). Metacognitive Strategies and Learning Quality: A Systematic Mapping Study. In a ResearchGate 98(2), *Proceeding* 7th *International Conference on Educational Technologies, February, 2020* (pp. 978-989) Colombia, Universidad de Córdoba. doi:10.33965/icedutech2020_202002L007

Rotter, J. B. (1960). Some Implications of a Social Learning Theory for the Prediction of Goal Directed Behavior from Testing Procedures. *Psychological Review* (Vol. 67), 301-316. Retrieve from https://psych.fullerton.edu/jmearns/rotter.htm

Ruth, G., & Patricia, A. A. (1989). Metacognition: Answered and Unanswered Questions. *Journal of Educational Psychologist, Taylor & Francis, 24(2).* doi:org/10.1207/s15326985ep2402_2

Schraw, G., & Moshman, D. (1995). Metacognitive Theories. *Educational Psychology Review, 7,* 351-371. doi.org/10.1007/BF02212307

Sergey, T., Olga, K., & Victor, T. (2015). Efficiency and Quality of University Performance: Faculty Opinion. *International Journal for Quality Research, 9*(3), 435-454. Retrieved from https://www.researchgate.net/publication/283878768

Sergey, V., Arkhipov, E. N., Vanchikova, N. A., Zolotareva, A. E., Yantranov, D. T., & Budaeva, (2019). Research into Motivational Factors of Work Done by

University Teachers from the Perspective of the Theory of Generations. *TEM Journal, 8*(4), 1477-14183. doi: 10.18421/TEM84-53

Sharul, E. J., Ruhanita, M., Amizawati, M. A., & Nor, L. A. (2015). Performance Measurement System and Lecturers' Performance: Testing the Mediation Role of Competency in Malaysian Research Universities. *International Business Education Journal, 8*(1), 105-120. Retrieved from https://www.researchgate.net/publication/303822759

Shaoning, Q. (2018). Exploration and Research on Educational Governance in Higher Vocational Colleges. *Education and Humanities Research, Atlantis Press*, (Vol.300). doi: 10.2991/erss-18.2019.59

Siti, R. A. H., Sharifah, S. S. H., & Nik, A. H. I. (2012). Teaching Quality and Performance among Experienced Teachers in Malaysia. *Australian Journal of Teacher Education,ResearchGate, 37*(11). doi:10.14221/ajte.2012v37n11.2

Smaragda, K., Elena, K., Nikolaos, M., George, S., & Andreas, D. (2019). Cognizance in Cognitive Development: A Longitudinal Study. Elsevier, (Vol.52). Retrieved from www.elsevier.com/locate/cogdev

Sonia, O. B., & Sandra, J. M. (2008). Metacognicion: Un Camin Para Aprender A Aprender, Metacognititon: a Way Towards Learning how to Learn. *Estudios Pedagógicos* (Vol. 1), 187-197. Retrieved from https://scielo.conicyt.cl/pdf/estped/v34n1/art11.pdf

Stanislav, S., Donatas, G., Nele, Z., & Tomas, K. (2025). Metacognitive Strategies Improve Self-Regulation Skills in Expert Sports Coaches. *Scientific Report, Nature Portfolio,* 15-34. doi.org/10.1038/s41598-025-86606-7

Stephanie, P., Elmar, S., & Rainer, B. (2013). Adaptation to Context as Core Component of Self-Regulated Learning: The Example of Complexity and Epistemic Beliefs. *Springer, New York, 28,* 53-65. Retrieved from http://hdl.handle.net/1959.13/1356951

Steve, H. (2020). Quantitative Analysis of Questionnaires, Techniques to Explore Structures and Relationships. *Routledge, Tylor & Francis Group, London and New York.*

Sweller, J. (1988). Cognitive Load during Problem Solving Effects on Learning. *Science Direct, Cognitive Science, Elsevier, 12(2),* 257-285. doi.org/10.1016/0364-0213(88)90023-7

Tahseen, A., Ouyang, G., Muhammad, A. H., Jordi, C., Sumaira, K., & Noor, A. (2020). Moral Education for Sustainable Development: Comparison of

University Teachers' Perceptions in China and Pakistan. *Sustainability, 12*(7). Retrieved from https://doi.org/10.3390/su12073014

Thenmozhi, C. (2019). Models of Metacognition. Shanlax. *International Journal of Education, 7*(20), 1-4. Doi:10.34293/education.v7i2.303

Tina, J. H., & Mark, A. S. (2012). Teacher Planning, Instruction and Reflection: What We Know About Teacher Cognitive Processes. *Quest, 58*(4), 424-442. doi:10.1080/00336297.2006.10491892

Turgut, K., David, G., Tijen, T., & Sedat, K. (2025). What Factors Mediate the Relationship Between Leadership for Learning and Teacher Professional Development? Evidence from Meta-analytic Structural Equation Modelling. *Educational Management Administration and Leadership, Sage,* 1-24. doi: 10.1177/17411432241308461

Valerie, E. L., Robert, F. D., & Julia, B. S. (1991). The Effect of the Social Organization of Schools on Teacher's Efficacy and Satisfaction. Sociology of Education, *64*(3), 190-208. doi:10.2307/2112851

Vibha, K., & Christine, A. W. (2019). Pragmatism as a Research Paradigm and its Implications for Social Work Research. Social Sciences, (Vol. 8). doi:10.3390/socsci8090255

Vygotsky, L., S. (1934). *Myshlenie i rech' Thinking and speech. Moscow: Gosudarstvennoe sotsial'no-ekonomicheskoe izdatel'stvo*. Retrieve from https:// 6813fc2f-fb64-8006-ab8c-1a86bb121faa

Waltor, D. (2005). Classroom Management and Organization. *Hamilton.*

Weilin, Y. (2023). Improving Teaching Effectiveness in Higher Education. *ASSEHR,* (Vol.675), 1126–1134. Retrieved from https://www.atlantis-press.com/proceedings/seaa-22/125976920

Xiaoyu, J., Weijian, L., & Liren, C. (2019). The Role of Metacognitive Components in Creative Thinking. *Frontiers in Psychology,* (Vol.10). Retrieved from https://doi.org/10.3389/fpsyg.2019.02404

Xu, X., Yu, S. J., Pang, N., Dou, C. X., & Li, D. (2022). Review on A Big Data-Based Innovative Knowledge Teaching Evaluation System in Universities. *Journal of Innovation & Knowledge, 7*(3). doi:10.1016/j.jik.2022.100197

Appendix 1.

Conceptual Framework of the Study

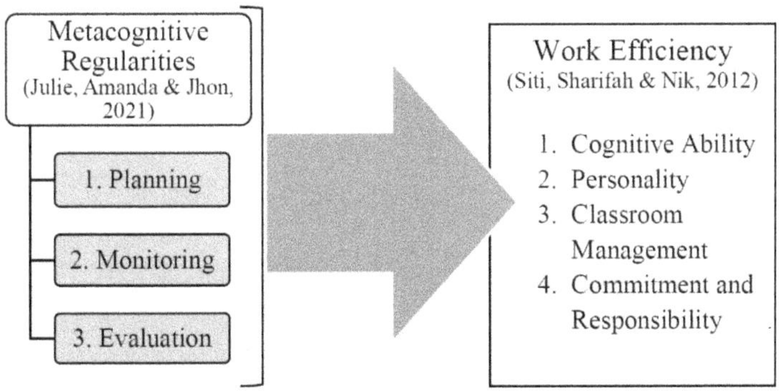

Appendix 2.

Operational Definition of the study

Metacognitive Regularities

Metacognitive regulation is referred to as a part of metacognition regulated by the process made up of three sub-components: planning, monitoring, and evaluation.

Planning

Organize work outcomes, align all areas with the institution's shared goals, and implement suitable approaches.

Monitoring

Monitoring involves evaluating progress and effectiveness of current procedures in completing responsibilities promptly.

Evaluation

Evaluation is a procedure for applicability, development, effectiveness, and impact. Control tactics simply analyze how successful instructional objectives are.

Work Efficiency

Efficiency in work saves time and money by planning and understanding tasks, working harder within a limited timeframe, and achieving goals.

Cognitive Ability

Cognitive ability refers to a person's ability to organize, recall, and gather information from their environment, including attention, memory, and reasoning, which is crucial for task completion.

Personality

Individuality is a distinct style of understanding, acting, and experiencing, characterized by natural and developed character patterns in interactions with others.

Classroom Management

Classroom management involves teaching strategies to create a conducive learning environment where performance standards are set for students.

Commitment and Responsibility

Work commitment is indicated by the employee's interest in assigned tasks, which are completed responsibly and on time.

Appendix 3.

The Questionnaire

Metacognitive Regularity Assessment Scale (MRAS)

Instruction:

Please mark your response against 1 to 5 that Indicate your response like (1= strongly disagree, 2= Disagree, 3= Neutral, 4= Agree, 5= Strongly Agree).

Sr. No	Code	Questions	SD 1	D 2	N 3	A 4	SA 5
		1. Planning It is a management process concludes by determining ways to enhance your effectiveness going forward in similar circumstances according to both internal and external input					
1	PL1	I do planning for the best learning environment.	1	2	3	4	5
2	PL2	I plan my pedagogical task a few days before for class.	1	2	3	4	5
3	PL3	I plan activities for my career development.	1	2	3	4	5
4	PL4	I feel lesson planning helps me in a better way.	1	2	3	4	5
5	PL5	I complete my commitment by using calendar.	1	2	3	4	5
6	PL6	I successfully conduct student activities by managing class time.	1	2	3	4	5
7	PL7	I feel planning helps me recognize teaching strengths.	1	2	3	4	5
8	PL8	I revise my responsibilities during lecture formation.	1	2	3	4	5
9	PL9	I can compensate for my weak point with proper plans.	1	2	3	4	5
10	PL10	I schedule my daily work activities in advance.	1	2	3	4	5

11	PL11	I plan course objectives to improve my course efficiency.	1	2	3	4	5
12	PL12	I always make target plans for exam topics.	1	2	3	4	5
		2. Monitoring The thoughtful checked of abilities and techniques to raise the potential of a preferred result can be learned in a manner that allows for transfer to new situations.					
	Code	**Questions**	**SD**	**D**	**N**	**A**	**SA**
13	M1	I keep track of my lesson plans to improve my lecture delivery.	1	2	3	4	5
14	M2	I observe the best way to teach topics is to do a discussion.	1	2	3	4	5
15	M3	I noticed that progress-tracked classes give the best grades in universities.	1	2	3	4	5
16	M4	I check student learning through presentations.	1	2	3	4	5
17	M5	I keep checking on my own pedagogies.	1	2	3	4	5
18	M6	I keep track of my instruction practice to help me with course responsibilities.	1	2	3	4	5
19	M7	I progress in my teachings through a series of achieved commitments.	1	2	3	4	5
20	M8	I believe impact monitoring helps to improve the curriculum.	1	2	3	4	5
21	M9	I agree that all class resources are updated by oversight.	1	2	3	4	5
22	M10	I always follow up on my university-assigned task.	1	2	3	4	5
		3. Evaluation An intense method of estimation usually involves connections between previous knowledge and new material, organizing and structuring content ideas, reorganizing one's cognitive processes, using schemes and a focus on proof and reasoning, as well as making connections between one's own experiences and realistic scenarios.					
	Code	**Questions**	**SD**	**D**	**N**	**A**	**SA**
23	E1	I consider course evaluation as an annual practice for the university.	1	2	3	4	5
24	E2	I think classroom observations are the key concern in teaching.	1	2	3	4	5

25	E3	I always improve my teaching thru valuation.	1	2	3	4	5
26	E4	I found gaps in my teaching through summing up.	1	2	3	4	5
27	E5	I detect students learning deficiencies by observation.	1	2	3	4	5
28	E6	I always evaluate my teaching task with commitment.	1	2	3	4	5
29	E7	I believe topic inspection promotes the subject's development.	1	2	3	4	5
30	E8	I always face policy challenges brilliantly at university.	1	2	3	4	5
31	E9	I know institutional standards established on valuation.	1	2	3	4	5

Work Efficiency Assessment scale (WEAS)

		1. Cognitive Ability A highly general mental capability that, among other things, comprises the capacity for logic, arrange, resolve issues, think creatively, understand complex things, learn fast, and gain insight is just how intellect or cognitive ability is defined.					
	Code	**Questions**	**SD**	**D**	**N**	**A**	**SA**
1	CA1	I use verbal reasoning in classroom teaching.	1	2	3	4	5
2	CA2	I utilize reasoning abilities for pedagogical tasks oversight.	1	2	3	4	5
3	CA3	I trust intellectual capacity promotes classroom instruction.	1	2	3	4	5
4	CA4	I utilize my working memory to avail career opportunities.	1	2	3	4	5
5	CA5	I examine any subject thru visual processing.	1	2	3	4	5
6	CA6	I take notice of conditions using my conceptual skills for task management.	1	2	3	4	5
7	CA7	I use social cognition to understand my students as a teacher.	1	2	3	4	5
8	CA8	I always solve issues using my thinking skills.	1	2	3	4	5
9	CA9	I utilize my intellectual capacities to take decisions for students.	1	2	3	4	5
		2. Personality					

		It is considered that compared to someone who is moderate, someone who is extreme on a specific feature level shows that characteristic's family of related behaviors more commonly. Thus, an illustration of a competitive attitude.					
	Code	**Questions**	**SD**	**D**	**N**	**A**	**SA**
10	P1	I am a very popular teacher of students.	1	2	3	4	5
11	P2	I think strong relationships with students and teaching go hand in hand.	1	2	3	4	5
12	P3	I am well aware that changing student behavior by strictness is not possible.	1	2	3	4	5
13	P4	I apply all of my qualities in teaching to help my students.	1	2	3	4	5
14	P5	I know behavior creates characters.	1	2	3	4	5
15	P6	I can control my mood swings in the classroom.	1	2	3	4	5
16	P7	I am able to control my feelings In difficult situations.	1	2	3	4	5
17	P8	I investigate the scenario before taking class decisions.	1	2	3	4	5
18	P9	I consider tracking makes it easy to advise students who are not active.	1	2	3	4	5
19	P10	I tackle my mission carefully.	1	2	3	4	5
20	P11	I am a versatile person for my colleagues.	1	2	3	4	5
		3. Classroom Management In this context, classroom management generally refers to the methods and techniques employed by teachers to address the issue of maintaining order in their classes.					
	Code	**Questions**	**SD**	**D**	**N**	**A**	**SA**
21	CM1	I manage silence in classroom for discipline.	1	2	3	4	5
22	CM2	I generally utilize a teamwork strategy in class.	1	2	3	4	5
23	CM3	I observe, It's ideal for students to sit in straight lines.	1	2	3	4	5
24	CM4	I focus on respecting university regulations in the classroom.	1	2	3	4	5

25	CM5	I like to design classrooms with a focus on the students.	1	2	3	4	5
26	CM6	I help pupils on a regular basis to motivate them.	1	2	3	4	5
27	CM7	I know planning creates excellent class time.	1	2	3	4	5
28	CM8	I utilize only the administration's provided resources in classroom.	1	2	3	4	5
29	CM9	I constantly ask students to keep the classroom tidy by observing them.	1	2	3	4	5
		4. Commitment and Responsibility A commitment is an intellectual or passionate attachment to a group's goals and ideals, to how one fits into those goals and values, and to the group as a whole.					
	Code	**Questions**	**SD**	**D**	**N**	**A**	**SA**
30	CR1	I am responsible for completing the subject curriculum.	1	2	3	4	5
31	CR2	I wisely meet the demands of the student's learning.	1	2	3	4	5
32	CR3	I try to engage every learner in my period.	1	2	3	4	5
33	CR4	I take part in all institution-wide events.	1	2	3	4	5
34	CR5	It is a teacher's responsibility to build up students' social skills.	1	2	3	4	5
35	CR6	I believe the duty of the teacher is to maintain academic standards.	1	2	3	4	5
36	CR7	I trust that the interpersonal abilities of students must be developed by teachers.	1	2	3	4	5
37	CR8	I always become a part of the university establishment workgroup	1	2	3	4	5
38	CR9	I spend my time updating university policies.	1	2	3	4	5
39	CR 10	I always offer a hand to other officials as a good university colleague.	1	2	3	4	5
40	CR 11	I give classroom evaluations with an accurate approach.	1	2	3	4	5

41	CR 12	I feel it is my responsibility to respect my seniors' comments heartily.	1	2	3	4	5
42	CR 13	I am committed to carrying out the classes at the appointed hour.	1	2	3	4	5

SEMI-STRUCTURED INTERVIEW SHEET

Section one Metacognitive Regularities

Planning

1. Do you feel planning comprises the correct vision for mission. How?

____

2. Do you think planning gives effect on student performance? Why?

____

3. Do you think planning improves course efficiency? How?

____

4. Do you complete commitment by using a calendar? How?

____

5. Do you schedule daily work activities in advance? Why?

____

Monitoring

6. Do you feel monitoring helps to organize the best work? How?

____

7. Do you think that monitoring is difficult to do? Why?

____

8. Do you think presentation helps student work monitoring? How?

____

9. Do you think monitoring helps in achieving step-by-step commitment to work? Why?

____

10. Do you agree that all class resources are updated by monitoring? How?

____

Evaluation

11. Does evaluation provide factual Support? How?

---

12. Does evaluation help to enhance the quality of work? Why?

---

13. Do you believe topic inspection promotes the subject's development? Why?

---

14. Do you think evaluation found gap in teaching process? How?

---

15. Do you consider course evaluation as an annual evaluation practice in an academic institution?

Why?

Metacognitive Regularity Effects on Work Efficiency

16. Do you agree that monitoring effects on teaching? Why?

---

17. Do you think that evaluation integrates the information? How?

Section two Work Efficiency

Cognitive Ability

1. Does Cognitive ability help in teaching? How?

---

2. Do you think that Cognitive ability helps in classroom management? Why?

Personality

3. Does the teacher's Personality affect the teaching and learning process? How?

____

4. Do you think Personality is a combination of attitudes? Why?

____

Classroom Management

5. Does the silence is effective for classroom management? Why?

____

6. Do the teachers use group activity known as classroom management? How?

____

Commitment and Responsibility

7. Do you feel that Effective performance is intervened by the teachers Commitment? How?

____

8. Do you think being committed increases your responsibility? Why?

____

Work Efficiency Determination

9. Does the university environment remain well-organized by the teachers? Why?

____

10. Do you accept that teacher affects students' overall well-being? How?

____

Index

www.ingramcontent.com/pod-product-compliance
Lightning Source LLC
LaVergne TN
LVHW010718210926
848485LV00008B/1451/J